Unless Recalled Earlier

DATE DUE

IMAGINING THE SCIENCES
Expressions of New Knowledge
in the
"Long" Eighteenth Century

IMAGINING THE SCIENCES
Expressions of New Knowledge in the "Long" Eighteenth Century

Robert C. Leitz, III
Kevin L. Cope
Editors

AMS PRESS
NEW YORK

LIBRARY OF CONGRESS CATALOGING-IN-PUBLICATION DATA

Imagining the Sciences: Expressions of New Knowledge in the
  "Long" Eighteenth Century / Edited by Robert C. Leitz, III and
  Kevin L. Cope.
     p. cm. – (AMS Studies in the Eighteenth Century; no. 43)
     Includes bibliographical references and index.
     ISBN 0-404-63543-1 (alk. paper)
     1. Science—History—17th century. 2. Science—History—
     18th century. 3. Science in literature. 4. Science in art.
     I. Leitz, Robert C., III, 1944-.   II. Cope, Kevin Lee.   III.
     Series
Q125.I393   2004
509'.032—dc21   2003048091
                                                          CIP

All AMS books are printed on acid-free paper that meets the guidelines
for performance and durability of the Committee on Production
Guidelines for Book Longevity of the Council on Library Resources.

AMS Press, Inc.
Brooklyn Navy Yard, Bldg. 292, Suite 417, 63 Flushing Ave.
Brooklyn, New York 11205
U.S.A.

Manufactured in the United States of America

CONTENTS

PART III: THE SCRIPTS OF SCIENCE

LIST OF ILLUSTRATIONS

FOREWORD

Robert C. Leitz, III and Kevin L. Cope

Louisiana State University in Shreveport and
Louisiana State University

"Oscillation" is one of a select group of otherwise ordinary English words that nowadays almost always appears in scientific contexts. Whatever dictionaries might opine, sound waves, cometary orbits, electrons, and pendula oscillate while, colloquially, people and opinions "vacillate" or "evolve" or simply "change." This subtle, ordinary-language distinction between cultural and scientific fluctuations epitomizes the complications that arise when scholars of the humanities, a discipline cast in the highly responsive media of language and the arts, confront science history, with its stiffer and harder vocabulary of planetary bodies, tectonic plates, incompressible minerals, and incontrovertible masses of experimental data. During the twentieth century, humanists' interpretations of science predictably oscillated between two extremes, with these extremes themselves oscillating between opposing varieties of "hardness" of softness." In the early to middle parts of the recently passed century, cultural historians of science like Richard Foster Jones or Marjorie Nicolson or Brian Vickers portrayed a semi-mythic "rise" of "new" science. Committed to the idea of "progress,"such an epic account of the birth and maturation of modern science drew attention to the confrontation between the humanities-trained intellectuals of the seventeenth century and the inundation of hard data that was pouring in through an assortment of new sources and technologies, whether oceanic travel or backyard telescopy. Not altogether unlike early twentieth-century heroic cinematic histories like *Birth of a Nation* or *The Ten Commandments*, this star-studded approach to science history juxtaposed an exciting but nevertheless hard and objective universe against a romantic picture of heroical scientists (Sir Francis Bacon, Sir Isaac Newton, even Benjamin Franklin) sailing like Wordsworth's Newton or Melville's Captain Ahab into unknown intellectual seas, all for the betterment—or, if the scientist happens to be mad, the worsening—of mankind. This romantic, optimistic approach persists to this moment in such pop-science hero-narratives as James Gleick's study of the recent history of non-linear mathematics, *Chaos*. More recently, under the influence of the contemporary interest in "cultural studies" and other social-historical approaches to literature, the history of science has moved to the opposite (if equally dyadic) extreme. Scholars of such diverse opinions as Douglas Lane Patey, Paul Korshin, Clement Hawes, and Barbara Shapiro have taken a more collectivist view of early science in which the emergence of a discourse or rhetoric or language or method or technique for natural-historical research was part of a complex cultural process to which a variety of disciplines, including literature and the arts, made contributions. For these interpreters, the language and institutions of science would never have emerged without the essayistic jottings of a Montaigne or a Bacon or the political ideas of a Locke or a Pufendorf, no more than the Newtonian interest in rainbows and refractions would have appeared without a contemporary interest in theatrical scene-painting or colorful special effects. This culturally attuned approach to early science, however, once again conceals an oscillation between the hard and the soft, the objective and the cultural. On one hand, the rhetorical position of these cultural historians is more "hard" than that of their predecessors, for it attempts to account for all the hard data about the cultures and times of early science, if only to avoid sentimentalizing a few celebrity scientists. On the other hand, this approach "softens" science by presenting it as if it were the same sort of cultural artifact as is an opera or painting—a position that ends up sentimentalizing nature by downplaying its intractability, by glossing over the hard fact that, say, a vulcanologist like Sir William Hamilton must deal with rocks and temperatures in a way from which Inigo Jones, when designing a set for a play on Pompeii, remains happily exempt.

In titling this book *Imagining the Sciences*, we have attempted to encourage a new approach to understanding science during the "long" eighteenth century and for that matter into our own day. To *imagine* the sciences is not exactly the same as to *chronicle* or even *explain* them. The latter requires making potentially exclusionary choices such as those made by historians of science to date: whether to focus retrospectively on individual heroes of science or on the mass of scientific culture; whether to regard science as an instrumental technology or a liberal art. To get past this theory-versus-practice dichotomy, *Imagining the Sciences* looks not so much at where science came from or at the steps in its rise in popular and academic prestige, but rather at the way in which the sciences as a whole were figured, prefigured, and refigured by early modern and Enlightenment practitioners, at the way that seventeenth- and eighteenth-century writers portrayed science and its promises, both to themselves and to the public. Although the volume abounds with interesting anecdotes and information from and about the early days of scientific inquiry, it is principally concerned with the imaging and understanding of the sciences themselves, with the way in which scientists saw science and concomitantly with the many colorful pursuits that resulted from this interest in science as a discipline, art, way of life, and hope for the future. Early on, science became a machine for producing not just science but colorful and imaginative science, for studying rainbows and rockets and rarities as much or more than the allegedly "useful" knowledge that supported the propaganda of the various national academies, "virtuosi," and "projectors." Although we have lost sight of its origins, this preference for colorful inquiries persists today in the difficulties encountered by those doing "basic" as opposed to special-project research and in the popular-culture affection for megaphenomena like black holes, time warps, extraterrestrial life forms, and cloned "Jurassic" dinosaurs. This book will look at the performance and the artifactual characteristics of early science, at, as the subtitle says, *expressions* of new knowledge in this era of expansive inquiry. "Expressions" here should not be confused with the "expression-ism" of out post-impressionist times. Rather, it should be understood in the sense that it carried in the Enlightenment: as the ventilation through widely available and often classically-sanctioned forms of ideas or passions, many of which ideas and passions were new to and unlicensed in the eighteenth cen-tury—or, more likely, licensed only through the enthusiasm for novelty and singularity that resulted from "imagining" the wonderful worlds of new science.

Imagining the Sciences is divided into three sections, each containing essays pertinent to this process of imagining an art of natural-historical inquiry. Section one, "The Theater of Science," addresses the capacity of early science to seek, prize, or even devise vast, grand, or otherwise impressive environments where the work of science can be done. In the opening essay, Bärbel Czennia acquaints us with an enormous space for scientific performance, the atmosphere, a place sufficiently large and dramatic for almost any kind of investigation, from the development of precision instrumentation to spectacularly dangerous exploratory balloon ascents. Czennia's multimedia article ranges across and re-integrates the seemingly disparate elements of early modern meteorology, from gimcrack equipment to sublime imaginations like those of Erasmus Darwin and Henri Fuseli. Barbara Benedict follows Czennia's contribution with an essay on mad scientists in the extended eighteenth century, an essay that traces the thin boundary between private eccentricity and public decorum, between the psychic energy that drives science and the hazardous products and performances that it produces. This first section concludes with Kevin Cope's plunge into deep interstellar, subterranean, and ultimately psychological space, specifically into

Henry More's intergalactic imaginations on an infinity of worlds, John Hutton's imaginative evocations of caverns and chasms, and Georg Friedrich Meier's probe into the unfathomable character of humor. For Cope, the paradox of "enlightened" science is its genial instability, its tendency to slip off the precipice of scientific vision into relentless affirmations of the infinite extensibility—and therefore ultimate unknowability—of the scientific universe.

The second section of *Imagining the Sciences*, "The *Dramatis Personæ* of Science," considers some representative instances of roles, poses, and lifestyles that the first scientists adopted. This section questions the relation between the scientific (or in some cases theoretical, metascientific) positions taken by investigators and the expressive modes through which those position are announced or lived out. Peter Walmsley inaugurates the section with a characterization of the imperial British scientist. Walmsley looks at the way that masculinity, scientific and technological competence, research, engineering, and colonialism all interact with one another. On the opposite side of Walmsley, this imperial coin is stamped Paul Johnston's "A God Who Must: Science and the Theological Imagination," which leads readers back from exotic British India to homespun colonial America, where fervid pulpit orator and committed Calvinist Jonathan Edwards finds that his God-given faculties compel him to study such out-of-the-way phenomena as spiders and thereby to work up both a scientific framework and scientific rhetoric in which to encapsulate his theological ideas. A similar link between extreme human experience and redemptive scientific study is proposed by Anna Battigelli, who connects Anne Conway's experience of chronic pain with the simultaneously materialist and mystical ideas of Francis Mercury van Helmont and other early explorers of energetic phenomena. The second section culminates with George Sebastian Rousseau's pursuit of the connections between dietary preferences (especially meat-eating) and stereotypes about national character. Rousseau's investigation of private, deeply personal issues such as the choice of one dish over another reveals the extent to which individual scientists were shaped, influenced, and nourished by national self-characterization and exposes the degree to which science runs on its belly, to which diet and menu subtly influence the chosen roles and intellectual behavior of scientific researchers even while they are providing topics for laboratory research.

The concluding third section, "The Scripts of Science," offers essays on the *process* of scientific life and research, especially on the interactive and narrative aspects of the scientific experience. Science in the "long eighteenth century" was increasingly a team effort. Along with the collectivization of research came an increased awareness of the way in which scientific undertakings were acted out along identifiable scripts, whether through real and imaginary interactions with other scientists or within the institutional characteristics of scientific culture or as constrained by the characteristics of the objective natural world. In "Cracks in the Cement of the Universe: Hume, Science, and Skepticism," Peter Fosl outlines the role of finitude in scientific research as perceived by Hume and other empiricists. Objects, observers, and the empiricist-touted "experience" were all subject to limitations and all forced by science into an interactive arena. Part of science was the recognition of and responding to the limitations, fragility, and general brittleness of all these players. The volume closes with James Buickerood's magisterial discussion of the role of thought experiments—the advance, preempirical scripting of scientific investigations—in Enlightenment science. Buickerood overturns conventional wisdom by rediscovering the role given to imaginary rehearsals of scientific tests and by drawing attention to the importance of nonempirical planning and outright fantasy speculation about experiences yet to come. For Buickerood, the theater of the imagination is as much of a draw for laboratory researchers as is the knowledge born of hands-on study.

The difficulty that eighteenth-century virtuosi experienced in trying to produce a vacuum suggests that idea-rich papers like those in *Imagining the Sciences* did not emerge from nullity. The foundational studies from which these enlarged and enhanced papers derived were initially presented at the first symposium-style conference of The Noel Collection of the Noel Memorial Library in Shreveport, Louisiana. In recent decades, economic, social, and professional factors have encouraged academic and independent researchers to convene and consult in large conferences where papers seldom exceed what Shakespeare might have regarded as the twenty minutes' traffic of the podium. Although this trend toward large meetings full of brief papers has greatly increased the audience appeal of eighteenth-century studies, it has come at the cost of the kind of deliberative and focused synods that were made famous in the middle of the twentieth century by Harvard's Villa I Tatti and other similar institutions. The "Science and the Imagination" conference that took place in November 2001 at The Noel Collection marks the first in a series of events that will draw together small ensembles of scholars working on specified themes and topics related to the seventeenth, eighteenth, and nineteenth centuries. In this way The Noel Collection hopes to encourage the use of its formidable resources, one strength of which is its holdings in early science, and likewise hopes to provide a venue for the scholarly treatment of topics that venture beyond safe, familiar, and mainstream scholarly concerns. Committed to an accessible and public discussion of the humanities, The Noel Collection conferences are and will remain open to the general as well as the scholarly public. They aim to bridge the gap between professional study of the eighteenth century and general-audience interest in and excitement over our Enlightenment cultural and intellectual heritage.

DEDICATION

In recognition of her commitment to and enduring
support of The Noel Collection and its research
programs, this book is gratefully dedicated to
Ruth Herring Noel.

ACKNOWLEDGMENTS

Recruitment and support of the scholars engaged in this project
required substantial resources. We would therefore like to thank
AEP/Southwestern Electric Power and also The Noel Foundation
for their generous support of this and many other initiatives of
The Noel Collection.

IMAGINING THE SCIENCES
Expressions of New
Knowledge
in the
"Long" Eighteenth Century

PART I
THE THEATER OF SCIENCE

FROM ÆOLUS TO ÆROLOGY,
OR,
BOREAS MEETS THE BAROMETER
Clouds, Winds,
and
Weather Observation in
Eighteenth-Century Poetry

Bärbel Czennia

Universität Göttingen

I: Atmospheric Research: The First Billows

I n the year 1798, chemist James Tytler introduces readers of
the first American edition of the *Encylopaedia Britannica* to an
exciting new branch of science:[1]

---

[1] *Encyclopaedia: Or, A Dictionary of Art, Sciences, And Miscellaneous Literature; Constructed on
a Plan, by which the Different Sciences and Arts Are digested into the Form of Distinct Treatises or
Systems, Comprehending the History, Theory, and Practice, of each, According to the Latest
Discoveries and Improvements...*, 18 vols. (Philadelphia, 1798). Printed and published with a
short dedicatory note by Thomas Dobson. For James Tytler's putative authorship of several
articles on natural sciences see "Preface," xiii.

AERO[RO]LOGY, The doctrine or science of AIR, its nature and different species, with their ingredients, properties, phenomena, and uses....Though the attention of philosophers has in all ages been engaged in some measure by inquiries concerning the nature of the atmosphere, yet till within these last 30 years, little more than the mere mechanical action of this fluid was discovered, with the existence of some anomalous and permanently elastic vapours, whose properties and relation to the air we breathe were almost entirely unknown. Within the abovementioned period, however, the discoveries concerning the constituent parts of the atmosphere itself, as well as the nature of the different permanently elastic fluids which go under the general name of *air*, have been so numerous and rapid, that they have at once raised this subject to the dignity of a *Science*, and now form a very considerable, as well as important, part of the modern system of natural philosophy.[2]

His characterization of the related discipline "Meteorology" is pervaded by the same awareness of imperfection and the necessity of further development:

[2] Vol. 1, 144–97, here 144. "Aerorology" in the title of the entry is obviously a typographical error, as the headlines of the following pages always read "Aerology." The length of the entry, extending over more than fifty pages, and the need for a special three-page index at its end, indicate the substantial increase of knowledge in this field during the last decades of the eighteenth century. This becomes particularly evident from a comparison with encyclopedic works published at the beginning of the century, like John Harris's famous *Lexicon Technicum*, 2 vols. (London, 1704–10): While Harris has shorter entries on "Air," "Atmosphere," "Clouds," "Meteors," "Wind" and "Vapours," he has none yet on either "Aerology" or "Meteorology." Ephraim Chambers's *Cyclopaedia, Or, An Universal Dictionary of Arts and Sciences*, 2 vols. (London, 1728), has an interesting two-page entry on "Weather" as well as articles on "Rain," "Heat," "Frost," "Hail," "Snow" etc. in the manner of Harris, and at least a two-line entry on "Meteorology." While the latter term is still missing in the first edition of the *Encyclopaedia Britannica*, 3 vols. (Edinburgh, 1768–71), short entries on "Aerography" and "Aerometry" make cross-references to a very long and thorough article on "Pneumatics," which contains an update of the latest research on air (including experiments with the air pump) as well as many meteorological aspects, like the distribution of winds around the globe.

In a science so very difficult, it is not to be supposed that any thing like a certain and established theory can be laid down: our outmost knowledge in this respect goes no farther as yet than to the establishment of a few facts; and in reasoning even from these, we are involved every moment in questions which seem scarcely within the compass of human wisdom to resolve.[3]

And Mr. Tytler is correct: Up to the seventeenth century the general understanding of weather was rudimentary and mainly based on conjectures derived from Aristotle's *Meteorologica*. This early attempt at a systematic description of atmospheric phenomena, however, turned out to be wrong in most points. Strictly speaking, meteorology did not become a science in the modern sense of the word before the nineteenth century and the subsequent arrival of refined mathematics, thermodynamics, and a better understanding of the relations between atmospheric air flow and air pressure. Improved instrument making and technological innovations like the electromagnetic telegraph then contributed to coordinating weather observation first on a national, later on a global scale.[4] Nevertheless, many important steps towards a new scientific meteorology had already been made in the course of the preceding two centuries:

---

[3] Vol. 6, 614. He reiterates this statement in a later entry on "Weather" (vol. 18, 820–35, here 821): "But meteorology is a science so exceedingly difficult, that, notwithstanding the united exertions of some of the first philosophers of the age, the phenomena of the weather are still very far from being completely understood; nor can we expect to see the veil removed, till accurate tables of observations have been obtained from every part of the world, till the atmosphere has been more completely analysed, and the chemical changes which take place in it ascertained." For the important role of encyclopedias for the promotion and popularization of scientific ideas see William Powell Jones, *The Rhetoric of Science. A Study of Scientific Ideas and Imagery in Eighteenth-Century English Poetry* (Berkeley and Los Angeles: University of California Press, 1966), 14–15.

[4] The first telegraph was installed in the United States in 1844, connecting Washington, D.C. with Baltimore. Joseph Henry's idea to equip the telegraph companies with meteorological instruments in exchange for a regular transmission of current weather-data to the Smithsonian Institute opened up new possibilities for weather-analysis and forecasting.

The seventeenth century saw the invention of scientific instruments like the mercury barometer and the thermometer, Robert Hooke's and Robert Boyle's pioneering experiments with the air pump, the first bloom of analytical chemistry, and the formation of the first scientific societies, creating new communication networks for the exchange of ideas as well as for the publication and general spreading of newly acquired knowledge. The eighteenth century witnessed major advances in the chemical analysis of air, the invention of Saussure's hygrometer, a better understanding of the links between air pressure, temperature, and atmospheric circulation, and an increasingly more systematic observation of weather phenomena like revolving storms, the formation of clouds, and various kinds of precipitation. Furthermore, the beginning of modern aviation, initiated by the Montgolfier brothers, opened up new possibilities of exploring the higher levels of the atmosphere.

The rapid accumulation of knowledge towards the end of the eighteenth century manifests itself in the growing number of entries on new scientific disciplines in contemporary encyclopedias, for which George Selby Howard's *New Royal Encyclopaedia Londinensis* is another good example. It promises

> Complete Systems...Digested into Distinct Treatises... Alphabetically Arranged, Critically Illustrated, and Copiously Explained, in the Most Clear and Satisfactory Manner...Including All the Latest Discoveries and Newest Improvements Made in the Various Branches of the Arts and Science.[5]

---

[5] *The New Royal Encyclopaedia Londinensis; Or, Complete, Modern and Universal Dictionary of Arts and Sciences*, 3 vols. (London, [1796?]). The quotation is from the title page. My conjecture with regard to the date of publication of this particular edition is based on the only specific date available, under the frontispiece to the left of the title page. The *English Short Title Catalogue* and the *British Library Public Catalogue* refer to an earlier edition, supposedly published in 1788 by the same publishing house (Alex. Hogg) under a slightly different title: *The New Royal Cyclopaedia, and Encyclopaedia; Or, Complete, Modern and Universal Dictionary of Arts and Sciences in Three Volumes...By George Selby Howard...Assisted by Many Gentlemen...John Bettesworth...Henry Boswell...Felix Stonehouse, D.D. and Others.*

The growing awareness of the complexity of atmospheric phenomena is also evident from the fact that articles on sub-disci-
plines with separate entries, like "Aerology," "Altimetry," "Anem-
ography," "Hydrography," or "Meteorology" are increasingly linked
by cross-references to related fields, like "Aerostation," "Chemistry," "Electricity," "Hydrostatics," "Magnetism," "Maritime
Affairs," "Navigation," "Physic" [sic] or "Pneumatics" in an
attempt to render as comprehensive an explanation of weather as
possible [see Figure 1].[6]

The progress of meteorology was the concern of much wider
circles than just the Royal Society and a handful of amateurs, at
a time when people were still much more exposed to and directly
dependent on weather than today.[7] This is also evident from the
enblematic frontispiece of Howard's *Encyclopaedia* [Figure 2]: in
the foreground of a neoclassical building resembling a temple filled
with scientific instruments, a female personification of knowledge
unveils globes and orreries to a male visitor, obviously representing
a modern scientist, or "natural philosopher" as he would still be
called. The background of the picture offers an outdoor panorama
dominated by a gigantic dark cloud, overshadowing both, a farmer
in the open field following a plough horse, and, somewhat further
in the distance, a ship under full sails approaching a southern shore
(indicated by two palm trees). Agriculture, navigation, and trade
were not only among the most important sectors of Britain's
eighteenth-century economy, but also particularly vulnerable to

[6] I would like to take this opportunity to express my gratitude to the James Smith Noel
Collection Curator, Prof. Dr. Robert C. Leitz, III, and the Noel Collection librarian, Martha
Lawler, for their gracious assistance and for permission to reproduce photographs of items
in their holdings. I would also like to thank the Noel Foundation for generous fellowship
support for my research.
[7] See the beginning of the entry on "Aerology" in Dobson, *Encyclopaedia*, vol. 1, 144: "those
discoveries, indeed, have not been more interesting to philosophers, than useful to science
and beneficial to society. In short...there is perhaps no station in life where some knowledge
of this subject may not be of use." For the new fashion among affluent English gentlemen
to set up private weather stations as well as for illustrations of contemporary equipment
see "Meteorological Instruments and Books" in Edward Morris, ed. (with contributions from
John Gage, Anne Lyles, Martin Suggett, John E. Thornes and Timothy Wilcox), *Constable's
Clouds. Painting and Cloud Studies by John Constable* (Edinburgh: National Galleries of
Scotland, Liverpool: National Museums and Galleries on Merseyside, 2000), 118–23.

the whimsies of weather, and therefore interested in practical applications derived from the progressing new branches of science. While many people were hoping for more reliable weather forecasts, some already dreamed of artificial weather manipulation on a global scale:

> The eastern parts of North America are much colder than the opposite coast of Europe, and fall short of the *stand-ard* by about 10° or 12°, as appears from American Meteorological Tables. The causes of this remarkable difference are many....It is covered with immense forests, and abounds with large swamps and morasses, which render it incapable of receiving any great degree of heat; so that the rigour of winter is much less tempered by the heat of the earth than in the old continent....But as the cultivated parts of North America are now much warmer than formerly, there is reason to expect that the climate will become still milder when the country is better cleared of woods, though perhaps it will never equal the temperature of the old continent....The winds evidently have a very great influence on the weather; the causes which produce them, therefore, ought to be examined with the greatest attention. Were we able to regulate their motions, we might, in a great measure, mould the climate of any country according to our pleasure; were we able to forsee them, it would be of the greatest importance to navigation and agriculture.[8]

---

[8] See Dobson, *Encyclopaedia,* on "Weather," vol. 18, 825–6.

II: Early Poetic Reverberations: Scientific Metaphors and Versified Meteorology in Seventeenth-Century Poetry

Speculations on and representations of the weather left their mark on English poetry long before the big upswing of atmospheric sciences towards the end of the eighteenth century.[9] The emerging scientific meteorology had a particularly strong impact on the imagination of poets writing since the time of the invention of the barometer. Henry More, fascinated with Galileo's theories on the rotation of the earth, speculates on its effects on the motion of air and on clouds in "A Platonick Song of the Soul" (c. 1647):

> In th' air they [the clouds] ly
> And whirl about with it, and when some wind
> With violence afore him makes them fly,
> Then in them double motion we find,
> Eastward they move, and whither by these blasts
> they're inclin'd.[10]

Andrew Marvell offers an interestingly de-centered perspective on condensation, evaporation, and the cycle of water in the atmosphere, in his ode "On a Drop of Dew" (1681):

> See how the Orient Dew,

---

[9] See Robert M. Schuler, *English Magical and Scientific Poems to 1700. An Annotated Bibliography* (New York, London: Garland, 1979), who examines the classical and native roots of the English tradition of scientific verse. With regard to weather-related poetry he refers, among others, to the weather lore in *The Georgicks of Hesiod* (that is *Works and Days*, translated into English by George Chapman in 1618), to Virgil's *Georgics*, and Lucretius's *De Rerum Natura* (also translated into English during the Renaissance period), as well as to English almanacs dating back to the fifteenth and sixteenth centuries, with various attempts at rhymed weather prognostication often connected to farming.

[10] Henry More, *A Platonick Song of the Soul*, ed. Alexander Jacob (Lewisburg: Bucknell University Press; London: Associated University Presses, 1998), 372, stanza 33, ll. 5–9.

Shed from the Bosom of the Morn
Into the blowing Roses,
Yet careless of its Mansion new;
For the clear Region where 'twas born
Round in its self incloses:
And in its little Globe's Extent,
Frames as it can its native Element.
How it the purple flow'r does slight,
Scarce touching where it lyes,
But gazing back upon the Skies,
Shines with a mournful Light;
Like its own Tear,
Because so long divided from the Sphear.
Restless it roules and unsecure,
Trembling lest it grow impure:
Till the warm Sun pitty it's Pain,
And to the Skies exhale it back again.[11]

Margaret Cavendish, temporarily intrigued with early modern reinterpretations of Epicurean atomism,[12] not only meditates on broad philosophical and epistemological questions like the reliability of the senses, or the possibility of a plurality of worlds,[13] but also on more specifically meteorological mysteries, like the origin of the winds, in her early collection *Poems and Fancies* (1653):

How can we thinke Winds come from Earth below,
When they from Skye do downe upon us blow?
If they proceeded from the Earth, must run

[11] From *Miscellaneous Poems*. See H. M. Margoliouth, ed., *The Poems and Letters of Andrew Marvell* (first ed. 1927), third ed. rev. by Pierre Legouis with the collaboration of E. E. Duncan-Jones (Oxford: Clarendon Press, 1971), 12–13, ll. 1–18.
[12] Developed by eminent philosophers like Descartes, Hobbes and Gassendi; see Anna Battigelli, *Margaret Cavendish and the Exiles of the Mind* (Lexington: University of Kentucky Press, 1998), 39–61. For her later conversion to vitalism see John Rogers, *The Matter of Revolution. Science, Poetry, and Politics in the Age of Milton* (Ithaca: Cornell University Press, 1996),187–88.
[13] See Battigelli, *Cavendish*, 52–53.

Strait up, and upon Earth againe backe come:…
Then sure it is, the Sun drawes Vapour out,
And rarifies it thin, then blow'th 't about.
If Heat condens'd, that turnes it into Raine,
And by its weight falls to the Earth againe.
Thus Moisture and the Sun do cause the Winds,
And not the Crudities in hollow Mines.[14]

As fanciful as they may appear today, Cavendish's scientific speculations in the medium of poetry show a particularly high degree of familiarity with contemporary scientific theories. It may be traced back to her acquaintance with early modern scientists whom she met during her exile in France. It is, for instance, remarkable that in the poem quoted above she dares to question the age-old belief in subterranean reservoirs of air as a cause of wind, otherwise still upheld by many learned men until the end of the following century. Further examples are her poems "*Of* Airy Atomes" and "*Of* Aire,"[15] which echo speculations on the shape of atoms laid out by Descartes in his essay *Les Météores* (1637) and in other contemporary renderings of corpuscular theory. Her poem "Aire begot of Heate and Moisture"[16] reveals ideas on the composition of air which could also be found in scientific tracts and later encyclopedias.[17]

Compared to poetry written in the course of the eighteenth century, however, empirical weather observation as a poetical subject is still a minor issue for most poets of this period. More's enthusiasm for representatives of scientific materialism (like

---

[14] "*Winds are made in the Aire, not in the Earth*," in *Poems and Fancies* (London, 1653, facs. repr. Menston, England: Scolar Press, 1972), 33. All subsequent quotations of poems by Margaret Cavendish follow this edition.

[15] *Poems and Fancies*, 7.

[16] *Poems and Fancies*, 33.

[17] See for example Dobson, *Encyclopaedia*, vol. 1, 147 and Howard, *New Royal Encyclopaedia*, vol. 1, 51, on the chemical properties of air. Another reason for the apparent modernity of Cavendish's poetical meditations on meteorological phenomena (in contrast to those of Marvell, for instance) lies in the absence of any discernable presence of God, which could be interpreted as religious skepticism. It is also typical of her later scientific reflections (see Battigelli, *Cavendish*, 101) and quickly invited accusations of atheism (Rogers, 194).

Copernicus and Galileo) was limited by his underlying neo-Platonist orientation. Marvell employs the dew drop's temporary separation and later return to the sky in order to make a statement on the human soul, which asks for a theological interpretation, rather than for a scientific explanation. As Anna Battigelli has convincingly argued, Cavendish used atomism primarily as "a metaphor" that helped her to explain "political and psychological conflict"[18] and less in order to promote the experimental philosophy of the forefathers of modern meteorology. On the contrary, some of her weather-related vignettes appear intentionally un- or even anti-scientific. The central agents in her poem on evaporation and condensation, for example, show a degree of erotic playfulness which is certainly incompatible with the treatment of the same topic in contemporary scientific tracts:

> Through *Earth's porous holes* her sweat doth passe,
> Which is the *Dew* that lyes upon the Grasse:
> Where (like a Lover kinde) the *Sun* wipes clean,
> That her faire face may to the Light be *seen*;
> And for her sake that *water* he esteemes,
> Threading those drops upon his silver *beames*,
> Like *ropes of Pearle*; he drawes them to his sphere,
> Turning those dropt to *Chrystall* when they're there....
> When she by sweat exhausted growes, and dry,
> The *Sun* the moystest *Clouds* doth squeeze in sky;
> Or else he takes some of his sharpest beames,
> To break the Clouds, from whence pure Chrystall streames.[19]

Cavendish's deliberate mixing of the latest speculations on the chemical composition of air with allegorical personifications borrowed from ancient mythology add up to an inconsistent, even

---

[18] Battigelli, *Cavendish*, 60. For Cavendish's use of contemporary scientific speculations as a metaphorical base for social and political issues see also Rogers, *Matter of Revolution*, 177–211, especially 181–90.

[19] *"Of the* Sunne, *and the* Earth," *Poems and Fancies*, 157, ll. 1–8 and 15–18.

contradictory statement. Her metaphysical conceits, defining chemical compounds and natural forces in terms of family relations or sexual intercourse,[20] can on one level be read as an attempt to reconcile the legacy of the past (represented by the Aristotelian organic tradition in science) with the radically different, atomistic and mechanistic concepts represented by early modern philosophers like Gassendi, Descartes and Hobbes.[21] Her change from apparently serious scientific speculations on the composition of air at the beginning of *Poems and Fancies*, to the sphere of superstition, legend, and folklore towards the end of the same volume, however, can also be interpreted as a provocation, revealing her unease with what she perceived as uncritical optimism towards the new empirical sciences:[22]

> *Lapland* is the place from whence all *Winds* come,
> From *Witches*, not from *Caves*, as doe think some;
> For they the *Aire* doe draw into high *Hills*,
> And beat them out againe by certaine *Mills*:
> Then sack it up, and sell it out for gaine
> To *Mariners*, which traffick on the *maine*.[23]

[20] Another example is her reference to ancient Roman wind deities in *"The* Windy Gyants," *Poems and Fancies*, 155–7, presented as "lustfull" and "so wilde, / As they doe force to get the Earth with childe," who then gives birth to an "Earthquake" (ll. 11–16).

[21] See Richard Johnson Sheehan and Denise Tillory, "Margaret Cavendish, Natural Philosopher: Negotiating between Metaphors of the Old and New Sciences,"*Eighteenth-Century Women. Studies in Their Lives, Work, and Culture* 1 (2001): 1–18.

[22] Following Richard Nate's argument in "'Poetical Cabbalas and Philosophical Worlds': Science and Literature in Margaret Cavendish's *Blazing World*" in Elmar Schenkel and Stefan Welz, eds., *Lost Worlds & Mad Elephants. Literature, Science and Technology 1700–1990* (Berlin: Galda + Wilch, 1999), 29–50, Cavendish's juxtaposition of scientific reasoning and fantastic fancy can be read as an expression of the "radical epistemological scepticism" towards natural philosophy "exhibited in her earlier works"(45), which led her "to equate philosophical theories with mere fantasies" (31). For Cavendish's increasingly more skeptical attitude towards Hobbesian materialism and the experimentalism represented by the Royal Society since the beginning of the 1660s (also expressed in the satire on Robert Boyle and Robert Hooke in her scientific fantasy story *The Description of a New Blazing World* [1666]), see Battigelli, *Cavendish*, 85–113.

[23] "Witches *of* Lapland," *Poems and Fancies*, 157.

The vacillation between attraction and rejection towards more recent models of meteorological explanation indicates the threshold position of a poet writing at a time when the old concept of science was no longer sufficient and the new one still bewildering. In the case of *Poems and Fancies*, the balance is restored by the mildly humorous undertone resulting from the comical incongruence of scientific reasoning and witch narrative forced together, albeit in separate chapters of the same book. The effect could be characterized as comic relief: At a time when the new sciences threatened to shatter age-old beliefs and ways to look at the world, with fundamental and often frightening implications for the fields of politics and theology, poetry provides a medium where the competing discourses can still be employed together in order to relate to the same topic (here: the origin of the winds) and yet create new meaning.[24] Exposing the growing incompatibility between ancient and modern science, science and folklore, becomes a new poetic theme in its own right and assigns to the sphere of fiction, especially to poetry, the role of a metadiscourse, where the bewildering contradictions can be explored in a delightful and entertaining manner.[25] To this end, Cavendish, in a highly self-conscious manner, puts various poetic sub-genres

[24] That even "non-sense" is regarded as a possible kind of meaning in poetry is made clear by the author in the rhymed postscript to her [unpaginated] introductory address "*To Naturall* Philosophers": "*Pray do not censure all you do not know, / But let my* Atomes *to the* Learned *go. / If you* judge, *and* understand *not, you may take / For* Non-sense *that which* learning Sense *will make. / But* I *may say, as* Some *have said before, / I'm not bound to fetch you* Wit *from* Natures Store."

[25] See once more her preface "*To Naturall* Philosophers": "I cannot say, I have not heard of *Atomes*, and *Figures*, and *Motion*, and *Matter*; but not throughly *reason'd* on: but if I do erre, it is no great matter; for my *Discourse* of them is not to be accounted *Authentick*: so if there be any thing worthy of noting, it is a good Chance; if not, there is no harm done, nor time lost....And the Reason why I write it in *Verse*, is, because I thought *Errours* might better pass there, then in *Prose*; since *Poets* write most *Fiction*, and *Fiction* is not given for *Truth*, but *Pastime*...I desire all that are not quick in apprehending, or will not trouble themselves with such small things as *Atomes*, to skip this part of my *Book*, and view the *other*, for feare these may seem *tedious*: yet the *Subject* is *light*, and the *Chapters* short. Perchance the *other* may please better...and if they cannot please, for lack of *Wit*, they may please in *Variety*, for most *Palates* are greedy after *Change*."

systematically to the test.[26] In between the extreme opposites of versified atomism and aerology on the one hand, and folk narrative[27] on the other, meteorological speculations are carried out in the form of the philosophical dialogue,[28] the love poem,[29] and even a meteorological "fancy." The latter, titled *"Similizing the Clouds to Horses,"* is another example for Cavendish's ability to reconcile contemporary scientific speculations on clouds and the dynamics of thunderstorms with a highly idiosyncratic poetic imagination:

The *Aiery Clouds* do swiftly run a *Race,*
And one another follow in a *Chase.*
Like *Horses,* some are sprightful, nimble, fleet,
Others sweld big with watry *Spavind Feet....*
They of all severall *Shapes,* and *Colours* be,
Of severall *Tempers,* seldom well agree.
As when we see *Horses,* which highly fed,
Do proudly snort, their *Eyes* look fiery red:
So *Clouds* exhaled, fed by the hot *Sun,*
With *Sulphur,* and *Salt-Peter* feirce become,...
Falling upon each others *Head,* and *Back,*
Never parted are, but by a *Thunder* Clap;
Pouring downe *Showres* of *Raine* upon the *Earth,*
Blow out strong *Gusts* of *Wind* with their long *Breath.*[30]

---

[26] This is indicated by the many different "prefaces" and other introductory explanations offered to different groups of potentially interested readers of the separate chapters.

[27] They are introduced by an [unpaginated] preface expressing her belief in the existence of "Fairies" and other supernatural beings.

[28] See *"A Complaint of Water, Earth, and Aire, against the Sun, by way of* Dialogue," *Poems and Fancies,* 61–3. It is part of a chapter consisting of nothing but versified "dialogues" and introduced with an address "To Morall Philosophers" (52–3).

[29] Unfortunately, the pagination of the book is very incoherent and contains several mistakes; after the first run from page 1 to 160, followed by four pages without numbers, the counting for "Poems. Of the Theam of Love" uses the page numbers 141 ff. for a second time. Weather-related poems in this chapter are the previously mentioned *"The* Windy Gyants" (155–57), "Witches of Lapland" (157), and the erotic reinterpretation of the natural cycle of water under the title *"Of the* Sunne, *and the* Earth" (157).

[30] *Poems and Fancies,* 142 (first use of this page number), ll. 1–4, 7–12, 17–20. Although "fancy" is not a poetic sub-genre on the same level as a versified philosophical dialogue or

Cavendish's subtle sense of "deep" humor is even more evident in the second part of the poem, offering a description of the final decomposition of the thunderstorm, where she further enriches her animated skyscape by the appearance of classical personifications of the west wind and the sun:

> Then *Boreas* whips them up, and makes them run
> Till their *Spirits* are spent, and *Breath* is gone;
> *Apollo* breakes, and backs them fit to ride,
> Bridling with his hot *Beames* their strengths to guide;
> And gives them *Heates*, until they foam, and sweat,
> Then wipes them dry, least they a Cold should get;
> Leades them to the middle *Region Stable*,
> Where are all sorts, dull, quick, weak, and able.   (ll. 21–28)[31]

III: The Poetic Atmosphere
from Anne Finch to
Erasmus Darwin

Anne Finch, "Upon the Hurricane." Empirical weather observations occupy a much more central position in Anne Finch's poetry written about fifty years later. In "Upon the Hurricane," written after the great storm that devastated Britain in November 1703, all the important ingredients of rough weather, high winds, rain clouds, thunder and lightning, are present and repeatedly referred to. The full title of her poem, "A Pindarick Poem. *Upon the*

a love poem, the introductory preface addressed "To Poets" and the character of the other texts assembled in this chapter suggest that a particularly unrestrained employment of fantasy is its main feature. The chapter as a whole gives the appearance of a mental exercise, displaying the poet's fantastic creativity and demonstrating her imaginative potential to the reader.

[31] The use of comic relief in Margaret Cavendish's versified scientific speculations seems to have gone unnoticed in most studies on her early work. It is not to be confused with her use of satire and parody in later works like *The New Blazing World*, for which Richard Nate, "Poetical Cabbalas" has coined the term "emotional relief" (35).

Hurricane in November 1703, *referring to this Text in* Psalm 148. ver. 8. Winds and Storms fulfilling his Word," suggests a fundamental difference in comparison with her female predecessor from the seventeenth century.[32] In contrast to Cavendish, Finch chooses an explicit and unbroken Christian orientation as her point of departure. Right from the start, her search for the cause of aerial motion points towards a personalized, metaphysical power rather than to the more abstract imbalance of air pressures:

YOU have obey'd, you WINDS, that must fulfill
The Great Disposer's righteous Will... (ll.1–2)

When you, fierce Winds, such Desolations wrought,
When you from out his Stores the Great Commander
    brought... (ll. 149–150)

Whilst You obey'd, you Winds! That must fulfill
The just Disposer's Righteous Will. (ll. 243–44)

On one level, Finch, then, interprets the hurricane of 1703 as storms and other "anomalies of nature" had been interpreted inside and outside English poetry for hundreds of years: as a punishment by an Old Testament God, notorious for his uncompromising severity towards his unruly believers, and well familiar from the Book of Psalms:[33]

Praise the LORD from the earth, ye dragons, and all deeps:

---

[32] This and all subsequent quotations from Anne Finch's poem follow the edition of the text given in *Miscellany Poems, on Several Occasions* (London, 1713), 230–41. It is a representative example of the long English tradition of versified biblical paraphrases; for the enrichment of this tradition by seventeenth- and eighteenth-century scientific thought, see Jones, *Rhetoric*, 40–54.

[33] Finch's conviction was shared by many early modern scientists, including Robert Boyle; see Rogers, *Matter of Revolution*, 193.

> Fire, and hail; snow and vapour; stormy wind fulfilling his word:[34]

Having pointed out various kinds of sinful behavior on the part of both, British individuals and the collective body politic,[35] her reflection on the vanity of human endeavors towards the end of the poem sounds like a tailor-made application of Psalm 148 for Britain at a time of impending political destabilization:[36]

> Thou far Renown'd, thou pow'rful BRITISH Isle!
> Foremost in Naval Strength, and Sov'reign of the Sea!...
> What art Thou, envy'd *Greatness*, at the best,
> In thy deluding Splendors drest?
> What are thy glorious Titles, and thy Forms?
> Which cannot give Security or Rest
> To favour'd Men, or Kingdoms that contest
> With Popular Assaults, or Providential Storms!
> Whilst on th' Omnipotent our Fate depends,
> And They are only safe, whom He alone defends.
> Then let to Heaven our general Praise be sent,
> Which did our farther Loss, our total Wreck prevent.
>   (ll. 285–298)

---

[34] The Holy Bible, Book of Psalms, 148:7–8. This and all subsequent Bible quotations follow a modern edition of the King James Authorized Version (Cambridge: Cambridge University Press, n.d.). In the 1713 edition Finch's poem is immediately followed by a versified paraphrase of this psalm. For the use of meteorological phenomena to illustrate the infinite power of an almighty God in the Book of Psalms see also the following passages: "He bowed the heavens also, and came down: and darkness *was* under his feet. / And he rode upon a cherub, and did fly: yea, he did fly upon the wings of the wind. / He made darkness his secret place; his pavilion [sic] round about him *were* dark waters *and* thick clouds of the skies."(Psalms, 18:9–11); "O Lord my God.../.../ Who coverest *thyself* with light as *with* a garment.../.../ ...who maketh the clouds his chariot: who walketh upon the wings of the wind:" (Psalms, 104:1–3); "He causeth the vapours to ascend from the ends of the earth; he maketh lightnings for the rain; / he bringeth the wind out of his treasuries." (Psalms, 135:7).
[35] See her comments on human pride and vanity (ll. 56–59, 66, 72,178, 259, 284–85, 289–93), impiousness (ll. 96–108), need for chastisement (ll. 110, 119, 190), greed for political power (ll.125–26, 175–86, 204–05), personal vices (ll.151–74), especially avarice (ll.162–74,192–99).
[36] See the implicit allusion to the War of the Spanish Succession which had already begun (l. 253), and the parallel drawn between contemporary Britain and ancient Judea (ll. 51–57).

But Finch's theological reading of the storm as a heavenly punishment does not keep her from also registering a remarkable amount of visually and acoustically perceptible activities on the lower levels of the atmosphere. In the opening lines she addresses the winds and directs the readers' attention to the relation between aerial motion and condensation, reflecting the meteorological understanding of her time:

> Throughout the Land, unlimited you flew,
> Nor sought, as heretofore, with Friendly Aid
>   Only, new Motion to bestow
> Upon the sluggish Vapours, bred below,
> Condensing into Mists, and melancholy Shade. (ll. 3–7)

The modern, empirical quality of Finch's detailed weather observation may easily be overlooked, owing to her retention of more traditional poetical devices; frequent use of anthropomorphism, for instance, lends a human face to inanimate components of nature in her minute description of the unusually scattered pattern of precipitation accompanying this particular storm:

> Ye *Clouds*! that pity'd our Distress,
> And by your pacifying Showers
> (The soft and usual methods of Success)
> Kindly assay'd to make this Tempest less;...
> In vain you wept o'er those destructive Hours,
> In which the Winds full Tyranny enjoy'd,
> Nor wou'd allow you to prevail,
> But drove your scorn'd, and scatter'd Tears to wail
> The Land that lay destroy'd. (ll. 233–242)

It also adds a dynamic quality to her observation of an unusual occurrence of the sun shortly before this storm:

> Which to foreshew, the still portentous *Sun*
> Beamless, and pale of late, his Race begun,

Quenching the Rays, he had no Joy to keep,
In the obscure, and sadly threaten'd Deep.   (ll. 247–250)[37]

Her attempt at an adequate characterization of the impressive noise of the wind illustrates the sensitivity of her "list'ning sense" (l. 146) for meteorological detail. In order to draw attention to the acoustic dimension of the storm, she offers expanded similes relating to sounds contemporary readers would be well familiar with; at the same time, they strengthen the central theme of never-ending human conflict and the resulting permanent military threat:

And in the loud tumultuous Jar
Winds their own Fifes, and Clarions are.
Each cavity, which Art or Nature leaves,
Their Inspiration hastily receives;
Whence, from their various Forms and Size,
As various Symphonies arise,
Their Trumpet ev'ry hollow Tube is made,
And, when more solid Bodies they invade,...
The beaten Flatt, whilst it repels the Noise,
Resembles but with more outrageous Voice
The Soldier's threatning Drum: (ll. 130–141)[38]

The accumulative evidence of the high number of weather agents repeatedly addressed and described in great detail, together with the occasional interspersion of terms borrowed from the new scientific discourse,[39] suggest Anne Finch's growing, albeit

[37] The passage is related to empirical weather data from 1703, which Finch considered interesting enough to be integrated verbatim into the poem, in the form of an explanatory note to line 247: "One Day of the Summer before the Storm, we had an unusual Appearance of the Sun (which was observ'd by many People in several Parts of Kent). It was of a pale dead Colour, without any Beams or Brightness for some Hours in the Morning, altho' obstructed by no Clouds; for the Sky was clear."
[38] For further allusions to the military theme see ll. 9–10, 81, 147, 150, 253, 293–99.
[39] See "Motion" (l. 50), "Condensing" (l. 7), "minuter parts of Air" (l. 183), "Number," "Weight" (l. 184), and the meteorological specification "HURRICANE" (l. 280).

involuntary or semi-conscious fascination with her subject: While coming close to running counter to her theological presuppositions (presenting the storm as a frightening act of divine justice), this shift of perspective prepares the ground for later, more enthusiastic representations of violent weather phenomena, for instance in the poetry of Ambrose Philips or James Thomson.[40]

Another major concern of Anne Finch is the impact of rain clouds and wind on the inhabitants of the devastated area.[41] She shares this interest in the human response to adverse weather with Jonathan Swift, who published his satirical "Description of a City Shower" only six years later, in 1710. For Swift, however, anticipating Pope's later programmatic statement that the "proper study of Mankind is Man" himself,[42] the description of the social interaction of London types, like the chambermaid, the poet, the lawyer, the sempstress, or the beau, is much more important than the description of the actual rain shower itself. The latter functions as a catalyst in a moral rather than a scientific test case. Modern meteorological speculations, which the title of the poem seems to invite, are at best ridiculed in the opening lines where fashionable weather observation and forecasting appear reduced to the level of superstition:

Careful Observers may foretel the Hour
(By sure Prognosticks) when to dread a Show'r:
While Rain depends, the pensive Cat gives o'er
Her Frolicks, and pursues her Tail no more.
Returning Home at Night, you'll find the Sink
Strike your offended Sense with double Stink....

---

[40] Ambrose Philips's poem "A Winter Piece" (1709) starts out as a complaint about the deficiency of northern winter weather, before it gradually moves from a detailed description to its positive evaluation. For Thomson see my analysis below.

[41] See the reports on various sites of damage in rural and urban areas (ll. 59–111) as well as on the open sea.(ll. 243–88).

[42] See line 2 of the second epistle of Alexander Pope's "An Essay on Man" (written 1730–32, published 1733–34) in John Butt, ed., *The Poems of Alexander Pope. A One-Volume Edition of the Twickenham Text* (New Haven: Yale University Press, 1977; first ed. 1963), 501–47, here 516.

> A coming Show'r your shooting Corns presage,
> Old Aches throb, your hollow Tooth will rage.[43]   (ll. 1–10)

The comparison of the rain cloud to an urban drunkard tells more about urban manners than the causes of precipitation:

> A Sable Cloud athwart the Welkin flings,
> That swill'd more Liquor than it could contain,
> And like a Drunkard gives it up again.   (ll. 14–16)

Considering the higher level of attention given to a canonical male author like Swift (as compared to Anne Finch), one might even wonder whether Swift's skeptical attitude towards evolving modern science distracted people's attention from the meteorological theme for a while—at least, until James Thomson rediscovered it two decades later in *The Seasons*.[44]

James Thomson, The Seasons. In contrast to Finch, Thomson rarely interprets the meteorological phenomenon as a form of divine punishment, despite occasional references to the same biblical storm scenery.[45] Based on a physico-theological approach to nature shared by many early modern scientists from Newton to Boyle, storms and countless other aspects of weather, such as

---

[43] All quotations from this poem follow the rendering of the text in Robert A. Greenberg and William B. Piper, eds., *The Writings of Jonathan Swift* (New York and London: Norton, 1973), 518–20.

[44] James Thomson, *The Seasons* in J. Logie Robertson, ed., *The Complete Poetical Works of James Thomson* (London: Oxford University Press, 1908), 1–249. All subsequent quotations follow this edition. "Winter" was first published separately in 1726, "Summer" in 1727, "Spring" in 1728, and "Autumn" in 1730. The parts were first published together in 1730, alterations made by the author as late as 1746.

[45] Echoes of the Book of Psalms can be heard in "Winter," ll. 197–201: "All Nature reels: till Nature's King, who oft / Amid tempestuous darkness dwells alone, / And on the wings of the careering wind / Walks dreadfully serene, commands a calm; / Then straight air, sea, and earth are hushed at once." The poem offers minute descriptions of storms occurring at various times of the year and in any imaginable location, including a whirlwind in the desert ("Summer," ll. 960–77), a "typhon" [sic] at sea ("Summer," ll. 980–1025), and a thunderstorm in rural Britain ("Summer," ll. 1103–1222); see also the destruction of the harvest ("Autumn," ll. 311–59), the extended study of a winter-storm ("Winter," ll. 72–201), and, more specifically, of a snow-storm ("Winter," ll. 223–321).

clouds, rainbows, vapors, fogs, frost and heat, become sources of awe, wonder and admiration testifying to the greatness of a benevolent Creator:

Hail, Source of Being! Universal Soul
Of heaven and earth! Essential Presence, hail!
To thee I bend the knee; to thee my thoughts
Continual climb, who with a master hand
Hast the great whole into perfection touched.
    ("Spring," ll. 556–60)

Inspiring God! who, boundless spirit all
And unremitting energy, pervades,
Adjusts, sustains, and agitates the whole....
The informing Author in his works appears:
Chief, lovely Spring, in thee and thy soft scenes
The smiling God is seen. ("Spring," ll. 853–60)[46]

Neither does Thomson have any difficulty to integrate seemingly negative phenomena into his *Argument from Design*: His reverence to ancient beliefs in the account of the "poisonous eastern winds," for example, is immediately followed by the assertion that their damage is still the lesser evil when compared to the destructive potential of the westerly Atlantic storms, which the former keep away from British coasts in spring.[47]

---

[46] While the first quotation echoes the idea of "the great manufacturer," the second focuses on God as the first cause of motion. A more indirect allusion can be found in "Winter," ll. 106–10: "Nature! great parent! whose unceasing hand / Rolls round the Seasons of the changeful year, / How mighty, how majestic are thy works! / With what a pleasing dread they swell the soul. / That sees astonished, and astonished sings!" That the latter statement should not be confused with nineteenth-century pantheism is obvious from a related statement by William Wollaston, quoted in Jones, *Rhetoric*, 23.

[47] See "Spring," ll. 137–42. The same applies for the death of a shepherd in the snow storm; the blame is neither put on the victim himself (for example as a punishment for possible moral shortcomings) nor on an indifferent or imperfect God, but, in a surprising turn, linked to social injustice in a society where the "shameful variance betwixt man and man" ("Winter," l. 331) results in harsh and risky living conditions for underprivileged people. In order to familiarize his British audience with a more positive attitude towards violent aerial motions, Thomson also employs cultural relativism, for example in his description of less

But piety and enthusiasm do not prevent Thomson (or his persona) from an accurate observation of nature, which is far more detailed than that of Anne Finch. Whatever outdoor panorama he describes, sooner or later the observing eye will be arrested in the upper end of the picture. Sometimes the observer gets so caught up in aerial activities that he temporarily loses visual contact with the ground altogether. Descriptions of landscapes turn into pure skyscapes, anticipating in the medium of poetry developments which would characterize English landscape painting in the second half of the eighteenth century.[48] Very often the general movement in these descriptions leads from top to bottom, from sky to earth. A good example is the beginning of a typhoon at sea:

> Amid the heavens,
> Falsely serene, deep in a cloudy speck
> Compressed, the mighty tempest brooding dwells.
> Of no regard, save to the skilful eye,
> Fiery and foul, the small prognostic hangs
> Aloft, or on the promontory's brow
> Musters its force. A faint deceitful calm,
> A fluttering gale...
>   Then down at once
> Precipitant descends a mingled mass
> Of roaring winds and flame and rushing floods.
>   ("Summer," ll. 986–96)

civilized peoples like the "sons of Lapland" who live in harmony with nature and "enjoy their storms" ("Winter," l. 846).

[48] See reproductions of sketches and paintings by eighteenth-century artists like Joseph Wright of Derby, Thomas Kerrich, Alexander and John Robert Cozen, Thomas Girtin, Cornelius Varley, and others in Anne Lyles' article, "'That immense canopy': Studies of Sky and Cloud by British Artists c.1770–1860" in *Constable's Clouds*, 135–50, esp. 136–44. See also John E. Thornes's essay "Constable's Meteorological Understanding and his Painting of Skies" in the same collection (151–59), quoting Constable on the emotional value of the sky for a landscape painting. His statement that "It will be difficult to name a class of Landscape—in which the sky is not the 'key note'—the 'standard of scale'—and the chief 'Organ of sentiment'" (156), expressed in a letter to a friend in 1821, seems to be equally valid for Thomson's much earlier verbal skyscapes quoted above.

ALL the VARIOUS DETACHED PARTS of KNOWLEDGE;

SELECTED FROM THE LATEST PUBLICATIONS,

ALPHABETICALLY ARRANGED, CRITICALLY ILLUSTRATED, and COPIOUSLY EXPLAINED, in the most *CLEAR* and *SATISFACTORY MANNER,*

ACCORDING TO THE *NEWEST* AND MOST *RESPECTABLE AUTHORITIES.*

ANCIENT and MODERN LITERATURE,

Comprising a Complete, Regular, and General Courſe of

From the very Earlieſt Ages of the World, down to the Preſent Time.

Including all the Lateſt DISCOVERIES and Neweſt IMPROVEMENTS made in the various Branches of the ARTS and SCIENCES,

PARTICULARLY IN

Acoustics	Brachygraphy	Engineering	Heraldry	Mechanics	Oratory	Recreations
Aerology	Bromatology	Engraving	History	Mensuration	Ornithology	Religion
Aerostation	Building	Entomology	Horsemanship	Mercantile Business	Orthography	Rhetoric
Agriculture	Carpentry	Ethics	Husbandry	Merchandize	Painting	Rites and Ceremonies
Alchemy	Catoptrics	Farriery	Hydraulics	Metallurgy	Perspective	Sculpture
Algebra	Chemistry	Fencing	Hydrography	Metaphysics	Pharmacy	Series and Statics
Altimetry	Chronology	Financing	Hydrostatics	Meteorology	Philosophy	Statuary
Amphibiology	Commerce	Fluxions	Ichthyology	Microscopical Disco-	Phlebotomy	Stereometry
Analytics	Conchology	Fortification	Laws and Customs	veries	Physic	Surgery
Anatomy	Conics	Gardening	Levelling	Military Matters	Physiology	Surveying
Anemography	Cosmography	Gauging	Logic	Mineralogy	Philology	Tactics
Architecture	Criticism	Geography	Longimetry	Modelling	Phytology	Theatricals
Arithmetic	Designing	Geometry	Magnetism	Music	Planometry	Theology
Astrology	Dialling	Glass-making	Mangery	Mythology	Pneumatics	Trades and Arts
Astronomy	Dioptrics	Grammar	Manufactures	Navigation	Poetry	Trigonometry
Belles-Lettres	Drawing	Gunnery	Maritime Affairs	Natural History	Politics	Typography
Book-keeping	Electricity	Handicrafts	Masonry	Nautical Matters	Projectiles	Vegetation
Botany	Elocution	Harmonics	Mathematics	Optics	Pyrotechny	Zoology, &c. &c. &c.

The Whole Entirely freed from the many Errors, Obſcurities, and Superfluities of other Works of the Kind; and containing the whole Subſtance (as far as relates to the ARTS and SCIENCES) of CHAMBERS and REES's CYCLOPÆDIA,—The SCOTCH and ENGLISH ENCYCLOPÆDIAS,—The FRENCH ENCYCLOPÆDIA, and every other English, Scotch, and Foreign Work of Eminence at preſent extant on the Subject; as well as all the other RECENT IMPROVEMENTS and DISCOVERIES comprized in the PHILOSOPHICAL TRANSAC-TIONS and other NEW PUBLICATIONS on SCIENCE.

Figure 1: Detail, title page, *The New Royal Encyclopaedia Londinensis.* Courtesy of the Noel Collection, Louisiana State University in Shreveport.

Figure 2: Detail, frontispiece, *The New Royal Encyclopaedia Londinensis*.
According to the small print below the picture it was
"Drawn by Ryley...Engraved by Widnell."
Courtesy of the Noel Collection, Louisiana State University in Shreveport.

[114]

—Thick as the dews, which deck the morning flowers,
Or rain-drops twinkling in the sun-bright showers,
Fair Nymphs, emerging in pellucid bands,
Rise, as she turns, and whiten all the lands.   10

I. "Your buoyant troops on dimpling ocean tread,
Wafting the moist air from his oozy bed,
AQUATIC NYMPHS!—you lead with viewless march
The winged Vapours up the aerial arch,   15
On each broad cloud a thousand sails expand,
And steer the shadowy treasure o'er the land,

The winged vapours, l. 14. See additional note, No. XXV, on evaporation.

On each broad cloud, l. 15. The clouds confist of condenfed vapour, the particles of which are too fmall feparately to overcome the tenacity of the air, and which therefore do not defcend. They are in fuch fmall fpheres as to repel each other, that is, they are applied to each other by fuch very fmall furfaces, that the attraction of the particles of each drop to its own centre is greater than its attraction to the furface of the drop in its vicinity; every one has afferved with what difficulty fmall fpherules of quickfilver can be made to unite, owing to the fame caufe; and it is common to fee on riding through fhallow water on a clear day, numbers of very fmall fpheres of water as they are thrown from the horfe's feet run along the furface for many yards before they again unite with it. In many cafes thefe fpherules of water, which compofe clouds, are kept from uniting by a furplus of electric fluid; and fall in violent fhowers as foon as that is withdrawn from them, as in thunder ftorms. See note on Canto I. l. 554.

If in this ftate a cloud becomes frozen, it is torn to pieces in its defcent by the friction of the air, and falls in white flakes of fnow. Or thefe flakes are rounded by being rubbed

[115]

Through vernal skies the gathering drops diffuse,
Plunge in soft rains, or sink in silver dews.—

together by the winds, and by having their angles thawed off by the warmer air beneath as they defcend; and part of the water produced by thefe angles thus diffolved is abforbed into the body of the hailftone, as may be feen by holding a lump of fnow over a candle, and there becomes frozen into ice by the quantity of cold which the hailftone poffeffes, and beneath the freezing point, or which is produced by its quick evaporation in falling; and thus hailftones are often found of greater or lefs denfity according as they confift of a greater portion of fnow or ice. If hailftones confifted of the large drops of fhowers frozen in their defcent, they would confift of pure tranfparent ice.

As hail is only produced in fummer, and is alway attended with ftorms, fome philofophers have believed that the fudden departure of electricity from a cloud may effect fomething yet unknown in this phenomenon; but it may happen in fummer independent of electricity, becaufe aqueous vapour is then raifed higher in the atmofphere, whence it has farther to fall, and there is warmer air below for it to fall through.

Or, fink in filver dews, l. 18. During the coldnefs of the night the moifture before diffufed in the air is gradually precipitated, and as it fubfides adheres to the bodies it falls upon. Where the attraction of the body to the particles of water is greater than the attractions of thofe particles to each other, it becomes fpread upon their furfaces, or flides down them in actual contact; as on the broad parts of the blades of moift grafs: where the attraction of the furface to the water is lefs than the attraction of the particles of water to each other, the dew ftands in drops; as on the points and edges of grafs or grafs, where the furface prefented to the drop being fmall it attracts it fo little as but juft to fupport it without much changing its globular form: where there is no attraction between the vegetable furface and the dew drops, as on cabbage leaves, the drop does not come into contact with the leaf, but hangs over it repelled, and retains its natural form, compofed of the attraction and preffure of its own parts, and thence looks like quickfilver, reflecting light from both its furfaces. Nor is this owing to any oilinefs of the leaf, but fimply to the polifh of its furface, as a light needle may be laid on water in the fame manner without touching it; for as the attractive powers of polifhed furfaces are greater when in actual contact, fo the repulfive power is greater before contact.

Q 2

Figure 3: A representative example of Erasmus Darwin's stylistic and formal fusion of Augustan poetry and scientific treatise. Copy from the third edition (London, 1795). Courtesy of the Niedersächsische Staats- und Universitätsbibliothek, Göttingen, Germany.

Figure 4: Representations of aerial nymphs,
copied from ancient Roman gems, in Joseph Spence's *Polymetis*.
Courtesy of the Noel Collection, Louisiana State University in Shreveport.

Figure 5: Illustration of Erasmus Darwin's personification of "Tornado" for the third edition of *The Botanic Garden* (London, 1795). Courtesy of the Niedersächsische Staats- und Universitätsbibliothek, Göttingen, Germany.

The same technique is used to describe the beginning of a thunder-storm in Britain:

> Behold, slow-settling o'er the lurid grove
> Unusual darkness broods, and, growing, gains
> The full possession of the sky, surcharged
> With wrathful vapour, from the secret beds
> Where sleep the mineral generations drawn....
>   and in yon baleful cloud,
> A reddening gloom, a magazine of fate,
> Ferment; till, by the touch ethereal roused,
> The dash of clouds, or irritating war
> Of fighting winds, while all is calm below,
> They furious spring. ("Summer," ll. 1103–16)

Even where the visual data are complemented by information on acoustic effects and the reactions of animals or human beings, the primacy of the lower atmosphere as the main stage for the enactment of a meteorological drama remains unchallenged:

> When to the startled eye the sudden glance
> Appears far south, eruptive through the cloud,
> And, following shower, in explosion vast
> The thunder raises his tremendous voice.
> At first, heard solemn o'er the verge of heaven,
> The tempest growls; but as it nearer comes,
> And rolls its awful burden on the wind,
> The lightnings flash a larger curve, and more
> The noise astounds, till overhead a sheet
> Of livid flame discloses wide, then shuts
> And opens wider, shuts and opens still
> Expansive, wrapping ether in a blaze....
>   Wide-rent, the clouds
> Pour a whole flood; and yet, its flame unquenched,
> The unconquerable lightning struggles through,
> Ragged and fierce, or in red whirling balls,

And fires the mountains with redoubled rage.
   ("Summer," ll. 1129–49)[49]

Likewise, in the description of an approaching winter storm, the reader is confronted with a dramatic account of quickly changing cloud formations, culminating in a light refraction (halo) around the moon and shooting stars:

When from the pallid sky the Sun descends,
With many a spot, that o'er his glaring orb
Uncertain wonders, stained; red fiery streaks
Begin to flush around. The reeling clouds
Stagger with dizzy poise, as doubting yet
Which master to obey; while, rising slow,
Blank in the leaden-coloured east, the moon
Wears a wan cricle round her blunted horns.
Seen through the turbid, fluctuating air,
The stars obtuse emit a shivering ray;
Or frequent seem to shoot athwart the gloom,
And long behind them trail the whitening blaze.
   ("Winter," ll. 118–29)[50]

---

[49] Thomson's fascination with thunderstorm clouds is shared by William Gilpin, "On Landscape Painting, A Poem" (1792), ll. 429–39. This and all further references relate to the text in the second edition of William Gilpin, *Three Essays: On Picturesque Beauty; On Picturesque Travel; And On Sketching Landscape: To Which Is Added A Poem, On Landscape Painting* (London 1794), 91–143, here 114. Another example for the impact of Thomson's aerial vision on later eighteenth-century poets is Robert Bloomfield's "The Farmer's Boy": Written in 1798 and clearly influenced by Thomson's *Seasons* (see "Preface," iv, and the subdivision into four seasonal parts), its "Summer" section contains a detailed description of thunderstorm clouds at night (ll. 254–74). This and all further references relate to the text in the second edition, *The Farmer's Boy, A Rural Poem* (London, 1800).

[50] The parallel between progressing modern meteorology and an increased artistic interest in skyscapes towards the end of the eighteenth century is also evident from William Gilpin's interesting hybrid "On Landscape Painting" which, mediating between the sister arts of poetry and landscape painting, offers the following advice to young painters: "With studious eye examine next the vast / Etherial concave: mark each floating cloud; / It's form, it's colour; and what mass of shade / It gives the scene below, pregnant with change / Perpetual, from the morning's purple dawn, / Till the last glimmering ray of russet eve." (ll. 67–73); see also ll. 368–78: "The sky, whate'er it's hue, to landscape gives / A corresponding tinge.... /...Chuse thy sky; / But let that sky, whate'er the tint it takes, / O'er-rule thy

Accurate observation of meteorological facts which may appeal to a reader's scientific interest, and their translation into a language that stimulates the imagination often complement each other, as, for example, at the beginning of the winter storm:

> The keener tempests come: and, fuming dun
> From all the livid east or piercing north,
> Thick clouds ascend, in whose capacious womb
> A vapoury deluge lies, to snow congealed.
> Heavy they roll their fleecy world along,
> And the sky saddens with the gathered storm.
> Through the hushed air the whitening shower descends,
> At first thin-wavering; till at last the flakes
> Fall broad and wide and fast, dimming the day
> With a continual flow.   ("Winter," ll. 223–32)

The analogy between low hanging, voluminous clouds and pregnant sheep has a double function: for one, the simile is integrated with the imagined setting, one of Britain's countless hillsides, exposed and unsuitable for anything but sheep grazing, and therefore well suited to prepare the reader for the oncoming story of a shepherd's death in the cold. At the same time, the image illustrates the color scheme of the clouds, white at the top because still thermally active, and darker underneath, suggesting weight, as well as their visually perceivable growth rate and fluffy texture, indicating the unfinished process of condensation. It also illustrates the irregularity of their motion, their low dragging over high ground while carrying a cold (and therefore heavy) air mass. It finally illustrates their low horizontal speed, probably resulting from collision with hilltops in their way, causing a backlog and enforced upward movement, which in its turn intensifies the condensation. All of these factors taken together, add up to what Luke Howard roughly seventy years later in his pioneering

pallet" (in Gilpin, *Three Essays*, 112).

classification of clouds would identify as a specimen belonging to the cumulus family.[51]

Also apparent from the example is Thomson's talent for capturing a sense of process which is typical of most meteorological events, like thermal convection, the accumulation of fog, or the approach of storms and precipitation. His step-by-step technique[52] adds the characteristic dynamic quality to his landscapes and focuses on all kinds of changes: changes in color and illumination, in structure and texture, in speed or direction of movement. It is complemented by his strategy of repetition and variation, when he contrasts descriptions of summer storms with those of winter storms, or of a sky in the morning with a sky at night. No poet but Thomson would notice the fundamentally different appearance of spring clouds and autumn clouds with so much accuracy, before Luke Howard revolutionized meteorology in 1803:

> At least from Aries rolls the bounteous sun,
> And the bright Bull receives him. Then no more
> The expansive atmosphere is cramped with cold;
> But, full of life and vivifying soul,

[51] See Luke Howard, "On the Modifications of Clouds, and on the Principles of their Production, Suspension, and Destruction; being the Substance of an Essay read before the Askesian Society in the Session 1802–3" in Alexander Tilloch, ed., *The Philosophical Magazine* 16 (1803): 97–107 and 344–57, 17 (1803): 5–11, especially 16: 101 and 16: 353–57. Inversely, the scientist Howard enriches his database of cloud observation by references to poetical cloud descriptions, including references to Virgil's *Georgics* as well as an excerpt from Bloomfield's "The Farmer's Boy" (16: 102–04), which also compares wintry cumulus clouds to sheep: "There views the white-rob'd clouds in clusters driv'n, / And all the glorious pageantry of heav'n. / Low, on the utmost bound'ry of the sight, / The rising vapours catch the silver light; / These Fancy measures, as they parting fly, / Which first will throw its shadow on the eye, / Passing the source of light, and thence away, / Succeeded quick by brighter still than they. / For yet above these wafted clouds are seen / (In a remoter sky, still more serene,) / Others, detach'd in ranges through the air, / Spotless as snow, and countless as they're fair; / Scatter'd immensely wide from east to west, / The beauteous 'semblance of a *Flock* at rest. / These, to the raptur'd mind, aloud proclaim / Their MIGHTY SHEPHERD'S everlasting Name." See "Winter," ll. 246–62, in Bloomfield, *Farmer's Boy*, 91–92.

[52] The underlying rhetorical structure could be characterized as "At first...then...and ...until at last...." See also Michael G. Ketcham, "Scientific and Poetic Imagination in James Thomson's 'Poem Sacred to the Memory of Sir Isaac Newton,'" *Philological Quarterly* 61 (1982) 1: 33–50, here 44–45.

Lifts the light clouds sublime, and spreads them thin,
Fleecy, and white o'er all-surrounding heaven....
At first a dusky wreath they seem to rise,
Scarce staining ether; but by fast degrees,
In heaps on heaps the doubling vapour sails
Along the loaded sky, and mingling deep
Sits on the horizon round a settled gloom.
     ("Spring," ll. 26–31 and 147–51)

When the bright Virgin gives the beauteous days,
And Libra weighs in equal scales the year,...
          Attempered suns arise
Sweet-beamed, and shedding oft through lucid clouds
A pleasing calm...
          till the ruffled air
Falls from its poise, and gives the breeze to blow.
Rent is the fleecy mantle of the sky;
The clouds fly different.   ("Autumn," ll. 23–37)[53]

That Thomson's predominant interest in air and atmospheric
phenomena is linked to a serious interest in the evolving new
science of meteorology also becomes evident in the finishing lines
of "Autumn":

O Nature! all-sufficient! over all
Enrich me with the knowledge of thy works;
Snatch me to heaven; thy rolling wonders there,
World beyond world, in infinite extent
Profusely scattered o'er the blue immense,
Show me; their motions, periods, and their laws
Give me to scan.   ("Autumn," ll. 1352–58)

---

[53] For further examples of Thomson's interest in the processual character of meteorological events see the description of rain showers in "Spring," ll.155–91, followed by the formation of a rainbow, the description of the passage of an oceanic typhoon in "Summer," ll. 980–96, and the development of a thunderstorm in rural Britain in "Summer," ll.1102–49.

His language, despite its often hymnal tone, is nevertheless pervaded with a striking amount of early modern scientific terminology: Witness nouns like "atmosphere" ("Spring," l. 28, "Winter," l. 908), "mass," "atoms" ("Summer," ll. 289–90), "prognostic" ("Summer," l. 990), "causes and effects…/…Essence" ("Summer," ll. 1746–47), "motions, periods, and…laws" ("Autumn," l. 1357), "rising system" ("Autumn," l. 1361), "ether" ("Winter," l. 45 and l. 720),[54] "nitre" ("Winter," l. 694),[55] "illusive fluid" ("Winter," l. 716), verb forms like "condensed" ("Autumn," l. 707), "diffused" (in the sense of "widely dispersed") and "suffused" (in the sense of "deeply informed," "Winter," ll. 719 and 722), or adjectives like "microscopic" ("Summer," l. 288), "complex" ("Autumn," l. 1361), and "etheral" ("Winter," l. 694). Thomson repeatedly refers to meteorological hypotheses of the day concerning the relation between temperature, density of air and air pressure in the poem proper as well as in footnotes.[56] Central issues which atmospherical science was still unable to explain, like the origin of the winds (supposedly located near the pole caps of the earth), or the origin of frost, are also explicitly addressed:

> Ye too, ye winds! that now begin to blow…
> Where are your stores, ye powerful beings! say,
> Where your aerial magazines reserved
> To swell the brooding terrors of the storm?
> In what far-distant region of the sky,
> Hushed in deep silence, sleep you when 'tis calm?
>   ("Winter," ll. 111–17)

> What art thou, frost? and whence are thy keen stores
> Derived, thou secret all-invading power,

[54] In contemporary usage it referred to the matter of the highest regions of the atmosphere.

[55] Eighteenth-century scientists still believed that the ether contained various substances including an important acid which formed saltpetre [sic] or nitre. Thomson implicitly refers to ancient philosophy which held the ether to be igneous, and, owing to its kind influence upon the air, to be the cause of all vegetation.

[56] See "Spring," ll. 26–28, "Summer," ll. 308–17, and "Winter," ll. 694–97, for a scientific footnote for example his comment on line 641 (123).

Whom even the illusive fluid cannot fly?
Is not thy potent energy, unseen,
Myriads of little salts, or hooked, or shaped
Like double wedges, and diffused immense
Through water, earth, and ether?
　　("Winter," ll. 714–20)[57]

Representatives of the new "angel-winged" sciences, as he calls them ("Summer," l. 1740), who share Thomson's combined interest in the atmosphere with a (physico-) theological disposition are even received into his modern pantheon of British national worthies, so far mainly populated with eminent politicians, soldiers and poets.[58]

At the same time, Thomson's meteorologically accurate observation of clouds has a clearly aesthetic dimension. His interest in their texture, color and shape as a stimulant for the poetic imagination is apparent in the following passage quoted from "Summer":

The Sun has lost his rage: his downward orb
Shoots nothing now but animating warmth
And vital lustre; that with various ray,
Lights up the clouds, those beauteous robes of heaven.
Incessant rolled into romantic shapes,
The dream of waking fancy!
　　("Summer," ll. 1371–6)

His poetic attention to the dynamizing effect of cloud movement on a landscape reflects, maybe even encourages, the theoretical and

---

[57] See also his reflections on the origin of winter ("Winter," ll. 894–901).

[58] He explicitly mentions Newton and Boyle: "Here, awful Newton, the dissolving clouds / Form, fronting on the sun, thy showery prism…" ("Spring," ll. 208–09); "Why need I name thy Boyle, whose pious search, / Amid the dark recesses of his works, / The great Creator sought? /…/ Let Newton, pure intelligence, whom God / To mortals lent to trace his boundless works / From laws sublimely simple, speak thy fame / In all philosophy…" ("Summer," ll. 1556–63).

practical attention English eighteenth-century landscape painters pay to clouds:

> And the sudden sun
> By fits effulgent gilds the illumined field,
> And black by fits the shadows sweep along—
> A gaily chequered, heart-expanding view,
> Far as the circling eye can shoot around,
> Unbounded tossing in a flood of corn. ("Autumn," ll. 37–42)[59]

Thomson's "comprehensive celebration of the metaphysical and aesthetic virtues of the new science"[60] thereby prepares the way for later poets like Wordsworth and Shelley, whose aesthetic interest in clouds was likewise based on solid meteorological knowledge — much more solid than some later literary critics— in the interest of clear-cut literary periods and an attempt to dissociate "Romanticism" from its intellectual roots in the earlier eighteenth century— would admit.[61]

Erasmus Darwin, The Botanic Garden. If, with James Thomson, weather-related poetry takes a decidedly scientific turn, in order to point out the innovative character of Erasmus Darwin's work at

[59] The passage reoccurs nearly verbatim in Erasmus Darwin's poem *The Botanic Garden*, Part 1, *The Economy of Vegetation*: "So when light clouds on airy pinions sail / Flit the soft shadows o'er the waving vale; / Shade follows shade, as laughing Zephyrs drive, / And all the chequer'd landscape seems alive." (Canto II., ll. 607–10). It is echoed in Canto IV., ll. 7–8, where sylphs flying over the land cause similar lively patterns: "Quivering in air their painted plumes expand, / And coloured shadows dance upon the land." This and all further quotations from Darwin's poem follow the fourth edition, *The Botanic Garden, A Poem. In Two Parts. Part I. The Economy of Vegetation. Part II. The Loves of the Plants. With Philosophical Notes* (London, 1799). Part II. was written before Part I., and first published independently in 1789; Part I was written by 1791 and first published in 1792. In the interest of space I will abbreviate the title of Part I. as *EoV*, the title of Part II. as *LoP*.
[60] See Peter Gay, *The Enlightenment: An Interpretation*, 2 vols. (New York: Alfred Knopf: 1969); here vol. 2, 127.
[61] See, for example, William Wordsworth's poems "To the Clouds" (date of origin unknown, published 1842) and "Composed After a Journey Across the Hambleton Hills, Yorkshire" (written 1802, published 1807), or Percy Bysshe Shelley's "Ode to the West Wind" and "The Cloud" (both 1820), the latter of which J. E. Thornes, "Luke Howard's Influence on Art and Literature in the Early Nineteenth Century," *Weather* 39 / 8 (1984): 252-54, has shown to give a systematic illustration of Luke Howard's various cloud types.

the end of the same century, it could be said that an inverse movement takes place and "science takes a poetic turn." This holds true, although the poem is swarming with all kinds of fantastic creatures well familiar from the poetry of classical antiquity, but which, as Darwin explains in his introductory "Apology," are neither inappropriate nor merely decorative elements in a "philosophical poem":

> The Rosicrusian [sic] doctrine of Gnomes, Sylphs, Nymphs, and Salamanders, was thought to afford a proper machinery for a Botanic poem; as it is probable, that they were originally the names of hieroglyphic figures representing the elements. Many of the important operations of nature were shadowed or allegorized in the heathen mythology…The Egyptians were possessed of many discoveries in philosophy and chemistry before the invention of letters; these were then expressed in hieroglyphic paintings of men and animals; which after the discovery of the alphabet were described and animated by poets, and became first the deities of Egypt, and afterwards of Greece and Rome.[62]

What may at first sound like a conventional rhetorical gesture, like respect being payed to neoclassicist poetical genre-expectations by Darwin the poet, has much wider implications. Coming from the opposite field, it helps Darwin the scientist to demonstrate that the sphere of factual, empirically gained knowledge and the sphere

---

[62] "Apology," in *Botanic Garden*, xvii–xviii. See the continuation of the same thought in a footnote to *EoV*, Canto IV.: "From theses fables, which were probably taken from antient [sic] hieroglyphics, there is frequently reason to believe that the Egyptians possessed much chemical knowledge, which for want of alphabetical writing perished with their philosophers" (199). For Darwin's "very original, and modern, interpretation of myths and allegories" (as a special device in oral cultures for storing scientific knowledge) and for his systematic and allegorical use of ancient mythology in the poem, see Pierre Danchin's excellent essay, "Erasmus Darwin's Scientific and Poetic Purpose in the Botanic Garden," in Sergio Rossi, ed., *Science and Imagination in Eighteenth-Century British Culture / Scienza E Immaginazione Nella Cultura Inglese Del Settecento. Proceedings of the Conference Gargnano del Garda 12–16 April 1985* (Milano: Edizioni Unicopli, 1987) 133–50, here 145.

of the imagination have never been separate in the first place: If, with hindsight, it is possible to retrace the ancient wind gods and air sylphs to icons which in the symbolic language of a sunken high culture stood for "the elements" and other important agents in chemical processes, which are only now being reinterpreted and re-coded in the terms of modern chemistry, Darwin argues during the late eighteenth century, then science and poetry are no opposites at all and can be combined in new and unprecedented ways. The practical result of Darwin's reflections is an unusual and ambitious experiment which the author himself characterizes as an attempt

> to inlist Imagination under the banner of Science; and to lead her votaries from the looser analogies, which dress out the imagery of poetry, to the stricter ones, which form the ratiocination of philosophy.[63]

The first part of his synthesis of ancient and modern scientific thought, and of science and poetry, is *The Economy of Vegetation*. It consists of four cantos, loosely connected by a plot line built around a number of supernatural beings: Invited by the "Genius of the place" (that is the Botanic Garden) the Botanic Goddess descends from the air in her flying car[64] "in her character of presiding over the air" and "is received by Spring and the Elements."[65] She successively addresses the nymphs, gnomes and sylphs related to fire, earth, water and the air before she alights again and disappears into a fine-weather sky.

---

[63] *The Botanic Garden*, "Advertisement," iii.
[64] This lends her some striking resemblance with Joseph Spence's description of the Roman Goddess Juno in his *Polymetis: Or, An Enquiry Concerning the Agreement Between the Works of the Roman Poets, and the Remains of the Antient Artists. Being an Attempt to Illustrate Them Mutually From One Another. In Ten Books* (London, 1747). See "Book the Sixth. Dial. XIII. Of the Beings, supposed to inhabit the Air," 201–13, here 210: "Juno...is represented...in a light car, drawn by peacocks." (see also plate XXXIX, 1., 333). For further indications of Darwin's extensive use of Spence's comparative study on ancient poetry and art, see his explicit references in footnotes to *EoV*: 24, 45, 185, 213, as well as more implicit references in the "Apology," (xviii) and footnotes to *EoV*: 5, 35, 96, 156–57.
[65] "Argument of the First Canto," xix.

The inlistment of the imagination by science has radical consequences for the outward appearance of the text, which, on many pages looks more like an early modern doctoral dissertation than a poem, owing to the number of "philosophical notes" which frequently cover more than fifty percent of the printed pages[see Figure 3].[66] Most relevant to atmospherical science because addressed to the supernatural inhabitants of the air is Canto IV. Considering, that in his introductory "Advertisement" Darwin had expressed the intention to explain "the operation of the Elements, as far as they may be supposed to affect the growth of Vegetables" (iv) even the first couple of lines of its "Argument" bear witness to a much wider concern:

ADDRESS to the Sylphs. I. Trade-winds. Monsoons. N.E. and S.W. winds. Land and sea breezes. Irregular winds... II. Production of vital air from oxygene and light... III. 1. Syroc. Simoom. Tornado... 2. Fog. Contagion...IV. 1. Barometer. Air-pump... 2. Air-balloon of Mongolfier. Death of Rozier...V. Discoveries of Dr. Priestley. Evolutions and combinations of pure air...VI. Sea-balloons, or houses constructed to move under the sea. (179)[67]

[66] While Darwin is not the first poet of the scientific age to complement a philosophical poem with scientific footnotes, the ratio between the text of EoV and its commentary is certainly extreme: while each of the four cantos covers approximately sixty printed pages, the footnotes and thirty-nine extended longer notes in the appendix fill another two-hundred-and-forty pages in much smaller print, and thus far exceed the poem in quantity. The scientific apparatus provides detailed information on all the scientific phenomena discussed in the poem proper as well as cross-references to related disciplines. Footnotes related to EoV, Canto IV., for example, comment on subjects as diverse as the convection of air, Benjamin Franklin's lightning rods and the electric charge of clouds, condensation and the formation of clouds, technical innovations in the field of weather-related measuring instruments and Irish meteorologist Kirwan's new theory on the winds. The most interesting longer "Note XXXIII. – Winds" covers thirty printed pages (403–33), has its own summary (487–89), and could without further changes be used as an entry in an encyclopedia, similar to those described at the beginning of my article. For earlier eighteenth-century poems with scientific notes see Jones, Rhetoric, 43, 46, 61, 77, 79.

[67] Plant-related aspects of air like "Seeds suspended in their pods" (appear comparatively late (Canto IV., l. 363 ff.) and at first rather scattered, still interspersed with headings like "Stars discovered by Mr. Herschel," "Concentric strata of the earth" or curiosa like "The crocodile in its egg." Only the latter third of Canto IV. (l. 421 ff.) discusses air in relation with

The structural backbone of this canto is formed by fifteen addresses of the Botanic Goddess to the sylphs of the air, naming and explaining their various "responsibilities" in her domain. Before the Goddess begins, Darwin offers his readers a clue as to the visual appearance of the aerial creatures **[see Figure 4]**:

> So round the GODDESS, ere she speaks, on high
> Impatient SYLPHS in gawdy circlets fly;
> Quivering in air their painted plumes expand,
> And coloured shadows dance upon the land.   (ll. 5–8)[68]

In its fanciful, nearly humourous character their introduction seems to serve a conciliatory purpose, mediating between traditional genre expectations (towards a poem) and an unusual theme (modern meteorology). In other words, they provide Darwin with a poetically acceptable framework for the bewildering mass of hard scientific facts which then follow, confronting the reader with an encyclopedic account of the nature, properties, and effects of air. The description of the sylphs' task number one, to guide the

---

vegetation in a more consistent manner, including aspects like pollination by wind, spreading of plant diseases by air and the chemical impact of vegetation on the atmosphere.
[68] The idea for this image once again seems to come from Spence's *Polymetis*, who, after a thorough description of the male wind deities, Zephyr, Boreas, Auster and Eurus, continues: "ALL that I have yet mentioned are males: for fear therefore that you should imagine the Winds to be a people of one sex only...we will now...consider some of those female figures....These are so many aerial nymphs: what the Romans called...Aurae; and what we call, Sylphs....They are all light and airy; generally with long robes, and flying veils; of some lively colour, or other: and fluttering about, as diverting themselves in the light and pleasing element, assigned to them. In short, they are...a species, of sportive, happy beings, in themselves; and well-wishers to mankind" (207–08 and plate XXVII.2., 3. and 4., 333). Spence's use of the term "species" (in the sense of "a class composed of individuals having some common qualities or characteristics, frequently as a subdivision of a larger class or genus" (earliest documented usage in this sense: 1630; see *The Compact Oxford English Dictionary* [Oxford: Clarendon, 1989],1844) must have attracted multi-talented Darwin's scientific mind; it indicates how by the mid-eighteenth century the new scientific terminology affected the language even of an Oxford professor of literature, long before Darwin's grandson Charles made the word more famous. Spence's taxonomic approach to ancient mythology must also have appealed to the botanist in Erasmus Darwin, who dedicated his poem to the chief taxonomist of the day, "the celebrated Swedish naturalist, LINNEUS" ("Advertisement," iii).

different winds around the globe, for instance, provides a lesson on global thermodynamic processes (ll. 9–12); the description of task number two, to "form with chemic hands the airy surge," (l. 29) a lesson on the chemical composition of air.[69] Task number three, to keep control over destructive winds, is followed by a scientific profile of winds harmful for human endeavors;[70] task number four, to inspire as muses the research of modern chemists, physicists and aeronauts, by a detailed account of the history of meteorology and its measuring instruments as well as visionary anticipations of technological progress.

Even before the beginning of Canto IV, Darwin's phenomenological interest in weather-related issues, especially in clouds, is evident. The latter are recorded from different view-points in all four section of this poem, which would rather deserve to be called a "rhyming encyclopedia." Reflections on the sun's impact on thermal convection in the fire-canto (I.) are followed by an examination of the aesthetic potential of cloud shadows on the ground in the earth canto (II.); the water canto (III.) addresses condensation and evaporation and follows the path of a dewdrop through the great "cycle of water" up to cloud level and down again as a rain drop.[71] Even the last image in the canto on air (IV.), at the very end of the poem, focuses on clouds: Anticipating modern uses of thermal up-currents for aviation purposes, it suggests a gliding flight as the cause of the Botanic Goddess's departure, whose magic car is pulled by "wandring Zephyrs,"

---

[69] Darwin's comparison of chemical components of the air to lovers (ll. 31–8) is reminiscent of the sexual innuendos in Margaret Cavendish's conceits discussed above.

[70] See for instance Canto IV., ll. 71-78: "You seize TORNADO by his locks of mist, / Burst his dense clouds, his wheeling spires untwist; / Wide o'er the West when born on headlong gales, / Dark as meridian night, the Monster sails, / Howls high in air, and shakes his curled brow, / Lashing with serpent-train the waves below, / Whirls his black arm, the forked lightning flings, / And showers a deluge from his demon-wings." For the related famous illustration, designed by Henry Fuseli and engraved by William Blake for the third edition (London: J. Johnson, 1795), see Figure 5, for further comment on Fuseli's idiosyncratic interpretation of Darwin's verse (representing Tornado as two distinct combatants) see Robert N. Essick's *William Blake's Commercial Book Illustrations. A Catalogue and Study of the Plates Engraved by Blake after Designs by Other Artists* (Oxford: Clarendon Press, 1991) 48.

[71] See ll. 201 ff., once again strongly reminiscent of Cavendish's *Poems and Fancies*.

"buoy'd in air" and followed by her attendants piercing "the slow-sailing clouds" "in spiral rings" (ll. 659–70).

This remarkable feat of technological prophecy is not the only one in the course of the poem. In addition to being a medical doctor, a poet, a botanist, and an amateur meteorologist, Darwin had close links to the Birmingham scene of industrialists.[72] Darwin's fascination with every technological innovation related to the medium air, is clearly stronger than his interest in extended nature walks à la Thomson, and the sensual experience of the sky they offer. Celebrating the "invention of the steam engine for raising water by the pressure of the air in consequence of the condensation of steam,"[73] for instance, leads him to the enthusiastic praise of its inventor as the 'Aeolus of the scientific age':

> NYMPHS! YOU erewhile on simmering cauldrons play'd,
> And call'd delighted SAVERY to your aid;
> Bade round the youth explosive STEAM aspire,
> In gathering clouds, and wing'd the wave with fire;
> Bade with cold streams the quick expansion stop,
> And sunk the immense of vapour to a drop.–
> Press'd by the ponderous air the Piston falls
> Resistless, sliding through it's iron walls;...
> Next, in close cells of ribbed oak confined,
> Gale after gale, He crowds the struggling wind;
> The imprison'd storms through brazen nostrils roar,
> Fan the white flame, and fuse the sparkling ore.
> Here high in air the rising stream He pours
> To clay-built cisterns, or to lead-lined towers;
> Fresh through a thousand pipes the wave distils,
> And thirsty cities drink the exuberant rills.
>   (*EoV*, Canto I., ll. 253–74)

---

[72] See Maureen McNeil, "The Scientific Muse: The Poetry of Erasmus Darwin" in L. J. Jordanova, ed., *Languages of Nature. Critical Essays On Science and Literature* (London: Free Association Books, 1986), 164–203.

[73] See Canto I., footnote to l. 254.

The god-like stylization of Captain Savery as the ruler over the winds testifies to Darwin's still unshattered faith in the natural right of mankind to control and subdue nature, justified by the result, improved living conditions for human beings. The awe-inspiring wildness nature still possessed in Anne Finch's hurricane poem and Thomson's storm studies is lost in the process of its taming and defeat.[74] Man-made machine, on the other hand, takes on the features and power of a magic dragon with "roaring brazen nostrils," anticipating Charles Dickens's famous description of Coketown in his industrial novel *Hard Times* (1854).[75]

Apart from technological progress, the exploration and domestication of the air provides the new age with new heroes of mythic dimensions. Next to inventors and industrialists like Thomas Savery, Darwin introduces his readers to the practitioners of a new kind of activity, which in contemporary encyclopedias could be found under the heading "AEROSTATION": [76]

While Darwin's vision of steam-powered aircraft for civil and military usage would have to wait for more than another hundred years to be put into practice,[77] ballooning had already become a reality in 1783. Just as in the days of ancient Greece, so the scientific age produces fortunate as well as unfortunate heroes, and both types are equally important for the creation of the new myth. The vessel of "the intrepid Gaul," Montgolfier, "[j]ourneying on high" in his "silken castle" "bright as a meteor through the azure tides" is likened to the successful travels of the ship "Argo,

---

[74] Darwin's "crowded struggling winds" and "imprisoned storms" are clearly reminiscent of the caves of Aeolus in ancient Greek mythology.

[75] See especially the beginning of chapter five ("The Keynote"), where Dickens associates the "town of machinery" with an urban jungle of exotic creatures, including painted savages, endless serpents and mad elephants.

[76] "...a science newly introduced into the Encyclopaedia. The word, in its primitive sense, denotes the science of suspending weights in the air; but in its modern acceptation, it signifies *aerial navigation*, or the art of navigating through the atmosphere." Dobson, *Encyclopaedia*, vol.1, 198.

[77] See *EoV*, Canto I., ll. 289–96: "Soon shall thy arm, UNCONQUER'D STEAM! afar / Drag the slow barge, or drive the rapid car; / Or on wide-waving wings expanded bear / The flying-chariot through the fields of air. / —Fair crews triumphant, leaning from above, / Shall wave their fluttering kerchiefs as they move; / Or warrior-bands alarm the gaping crowd, / And armies shrink beneath the shadowy cloud."

rising from the southern main."[78] The fate of Rozier, whose balloon crashed in flames in 1785, is compared to the tragic fate of Icarus:

> Fair mounts the light balloon, by Zephyr driven,
> Parts the thin clouds, and sails along the heaven;
> Higher and yet higher the expanding bubble flies,
> Lights with quick flash, and bursts amid the skies,...
> Headlong He rushes through the affrighted Air
> With limbs distorted, and dishevel'd hair,
> Whirls round and round, the flying croud [sic] alarms,
> And DEATH receives him in his sable arms!...
> So erst with melting wax and loosen'd strings
> Sunk hapless ICARUS on unfaithful wings.
>   (*EoV*, Canto IV., ll. 149–70)

The new myth is further enhanced by Darwin's prophetic vision of future deeds which will be even greater than those of the ancient supermen: awarding both balloonists the new term of honor, "philosopher" (that is "scientist"),[79] the death of the first victim in a modern aviation accident appears redeemed in the broader context of a dawning space age:

> Rise, great MONGOLFIER! urge thy venturous flight
> High o'er the Moon's pale ice-reflected light;
> High o'er the pearly Star, whose beamy horn
> Hangs in the east, gay harbinger of morn;
> Leave the red eye of Mars on rapid wing,
> Jove's silver guards, and Saturn's crystal ring;...
> Shun with strong oars the Sun's attractive throne,

---

[78] See *LoP*, Canto II., ll. 25–66.

[79] See "calm Philosopher" for Montgolfier (*LoP*, Canto II., l. 41) and "philosopher of great talents and activity" for Rozier (by Darwin spelled "Rosiere," *EoV*, Canto IV., footnote to l. 148). Darwin thus distinguishes both men by a more sophisticated kind of curiosity from many other curious people of their time, aptly characterized in Barbara Benedict's *Curiosity: A Cultural History of Early Modern Inquiry* (Chicago: Chicago University Press, 2001).

The sparkling Zodiac, and the milky zone;...
For thee Cassiope her chair withdraws,
For thee the Bear retracts his shaggy paws;
High o'er the North thy golden orb shall roll,
And blaze eternal round the wondering pole.
          (LoP, Canto II., ll. 47–62)

Triggered by the exploration of the atmosphere in balloons and reinforced by the progress made in the chemical analysis of air, Darwin even anticipates a parallel development in the exploration of the underwater world:

The diving castles, roof'd with spheric glass,
Ribb'd with strong oak, and barr'd with bolts of brass,
Buoy'd with pure air shall endless tracks pursue,
And PRIESTLEY'S hand the vital flood renew.
          (EoV, Canto IV., ll. 209–12)

While the lively description of the imaginary submarine antici-pates nineteenth-century *science fiction* à la Jules Verne, it also testifies to the technological and economic foresight of Darwin the entrepreneur in the beginning industrial era. His explicit reference to the commercial potential of practical applications of the new science(s) for the British Empire also foreshadows modern interdependencies of scientific research, politics, and economics:

Then shall BRITANNIA rule the wealthy realms,
Which Ocean's wide insatiate wave o'erwhelms;
Confine in netted bowers his scaly flocks,
Part his blue plains, and people all his rocks.
          (EoV, Canto IV., ll. 213–16)

If a degree of monetary interest in the exploration of the air cannot be denied, it is worth noting, that this is clearly not Darwin's only one, and that other passages in the same poem offer at least two additional motivations. For one, a deeply idealistic interest in the improvement of the living conditions of mankind as a whole, going far beyond the geographical and political boundaries of the British Empire. A good example are Darwin's reflections on the possibilities of artificial weather manipulation through the shifting of icebergs:

There, NYMPHS! alight, array your dazzling powers,
With sudden march alarm the torpid Hours;
On ice-built isles expand a thousand sails,
Hinge the strong helms, and catch the frozen gales.
The winged rocks to feverish climates guide,
Where fainting Zephyrs pant upon the tide;...
While swarthy nations crowd the sultry coast,
Drink the fresh breeze, and hail the floating Frost,
NYMPHS! Veil'd in mist, the melting treasures steer,
And cool with arctic snows the tropic year.
(*EoV*, Canto I., ll. 527–44)[80]

The clearly philanthropic motivation behind this act is further emphasized in a footnote related to the same passage:

If the nations who inhabit this hemisphere of the globe, instead of destroying their seamen and exhausting their wealth in unnecessary wars, could be induced to unite their labours to navigate these immense masses of ice into the more southern oceans, two great advantages would result to mankind, the tropic countries would be much cooled by their solution, and our winters in this

---

[80] The idea of a man-made balancing of global temperature zones appears to be so closely intertwined with scientific reflections on recently discovered natural laws (especially those concerning the (im)balance of pressure, weight, or electric charge of fluids, including air which was still regarded as one) that the benevolent attempt at a "social balance" may well be immediately derived from science itself rather than from Shaftesbury's concept of a natural altruism. In its idealistic orientation it certainly refutes the stereotypical Marxist image of Darwin as a representative of a class-conscious "industrial bourgeoisie" indifferent to the fate of anybody outside his own social group, brought forward by critics like McNeil, "Scientific Muse," 175–83.

latitude would be rendered much milder for perhaps a
century or two, till the masses of ice [at the pole] became
again enormous. (*EoV*, 60)

The second nonmaterialistic motivation behind Darwin's prophe-
sies of technological progress is the inherent value of daring
speculation itself, especially at a time when the incompleteness of
scientific knowledge was generally admitted and regretted by
natural philosophers. Darwin explicitly addresses the issue in his
introductory "Apology":

> It may be proper here to apologize for many of the
> subsequent conjectures on some articles of natural
> philosophy, as not being supported by accurate investiga-
> tion or conclusive experiments. Extravagant theories
> however in those parts of philosophy, where our know-
> ledge is yet imperfect, are not without their use, as they
> encourage the execution of laborious experiments, or the
> investigation of ingenious deductions, to confirm or
> refute them. And since natural objects are allied to each
> other by many affinities, every kind of theoretic distribu-
> tion of them adds to our knowledge by developing some
> of their analogies. (xvii)

An interesting illustration of this idea is provided by Darwin's
speculations on the origin of the wind, which had already occupied
the imagination of his poetic predecessors, Cavendish and
Thomson. In a first step, Darwin carefully prepares the poetic
ground for any further speculation with a reminder of the
prevailing shortcomings of atmospherical science:

> Oh, SYLPHS! disclose in this inquiring age
> One GOLDEN SECRET to some favour'd sage;
> Grant the charm'd talisman, the chain, that binds,
> Or guides the changeful pinions of the winds!
>   (*EoV*, Canto IV., ll. 319–22)

Although weather observation in Darwin's time already suggested a close connection between the abrupt fall of the barometer and the beginning of a storm, its actual causes, the exact relations between atmospheric air flow, air motion, and pressure, were as yet undiscovered. With eighteenth-century meteorologists being unable to explain what looked like a sudden disappearance (and inexplicable late reappearance) of air, all kinds of theories on the potential destruction of air (for example in subterranean caves or at the pole caps of the earth) and on the reproduction of the atmosphere were circulating. The most fantastic of all is probably Darwin's version, proving the strength of his poetic imagination at least as well as his talent for productive scientific speculation:

> Castled on ice, beneath the circling Bear
> A vast CAMELION drinks and vomits air;
> O'er twelve degrees his ribs gigantic bend,
> And many a league his gasping jaws extend;
> Half-fish, beneath, his scaly volutes spread,
> And vegetable plumage crests his head;
> Huge fields of air his wrinkled skin receives,
> From panting gills, wide lungs, and waving leaves;
> Then with dread throes subsides his bloated form,
> His shriek the thunder, and his sigh the storm.
> Oft high in heaven the hissing Demon wins
> His towering course, upborne, on winnowing fins;
> Steers with expanded eye and gaping mouth,
> His mass enormous to the affrighted South;
> Spreads o'er the shuddering Line his shadowy limbs,
> And Frost and Famine follow as he swims.
> (*EoV*, Canto IV., ll. 333–48)[81]

---

[81] A footnote to l. 334 refers the reader to the extended note "No. XXXIII" in the appendix to the poem where the speculations on a potential "officina aeris, a shop where air is both manufactured and destroyed in the greatest abundance within the polar circles"(406) continue on a somewhat more sober level for more than twenty pages, elaborating on four different models of explanation (406–25), before the idea of the dragon is seriously reconsidered: "…as there appears greater difficulty in accounting for this change of wind from any other known causes, we may still suspect that there exists in the arctic and

While Cavendish had disclosed her unease with the new scientific discourse by confronting her readers with a juxtaposition of two incompatible answers, the first vaguely scientific, the second deliberately fanciful,[82] and Thomson limited himself to merely stating the open scientific question ("Winter," ll. 111–17), Darwin redefines a seemingly fanciful answer, developed in the medium of poetry, as a serious contribution to scientific speculation. The most deserving meteorologist in Darwin's opinion, worthy to receive the "Golden secret" would be a member of the Royal Irish Society, Mr. Kirwan, who was famous for his publications on long-term weather observation (based on ship logbooks) and his theory of the winds.[83] Noteworthy in this context as well as the poem in general is Darwin's transnational perspective which enables him to celebrate Irish and Scottish "philosophers"(like Kirwan and Watt) as much as French explorers (like Montgolfier, Rozier, and Romain), French and English chemists (like Lavoisier and Boyle) as well as American (Franklin), Italian (Torricelli), or German scientists (von Guericke).[84] Darwin shares this remarkable

antarctic circles a BEAR or DRAGON yet unknown to philosophers, which at times suddenly drinks up, and as suddenly at other times vomits out one-fifteenth part of the atmosphere: and hope that this or some future age will learn how to govern and domesticate a monster which might be rendered of such important service to mankind. (425).

[82] See my analysis above of her more scientific speculation on the origin of the winds, somewhere up in the atmosphere, but not in the earth (*Poems and Fancies*, 33), as opposed to the more fanciful suggestion, Lapland (*Poems and Fancies*, 157).

[83] The celebration of this scientist once again leads back to Darwin's favorite idea of nature improved by human control: "SYLPHS! round his [that is the wind-eater's] cloud-built couch your bands array, / And mould the Monster to your gentle sway; / Charm with soft tones, with tender touches check, / Bend to your golden yoke his willing neck, / With silver curb his yielding teeth restrain, / And give to KIRWAN's hand the silken rein. / – Pleased shall the Sage, the dragon-wings between, / Bend o'er discordant climes his eye serene, / With Lapland breezes cool Arabian vales, / And call to Hindostan antarctic gales, / Adorn with wreathed ears Kampschatca's brows, / And scatter roses on Zealandic snows, / Earth's wondering Zones the genial seasons share, / And nations hail him 'MONARCH OF THE AIR.'" (*EoV*, Canto IV., ll. 349–62).

[84] Called "de Guerick" by Darwin (*EoV*, 192). Apart from the "big names" quoted above, many other scientists from various European countries are referred to in footnotes throughout the poem. Nevertheless, British "philosophers" outnumber their foreign colleagues; among the more frequently mentioned are Priestley, Cavendish, Bacon, Harvey, Newton, Herschel, and Halley.

lack of xenophobia (at a time otherwise characterized by the polemical rhetoric of nation building) with the scientific encyclopedia of his time. In contrast to Thomson, the poetical propagator of a distinctive British culture,[85] Darwin conceived of himself primarily as a member of the internationally minded community of modern scientists. More clearly than any of his predecessors, Darwin demonstrates that eighteenth-century poetry not only reflected the contemporary scientific discourse, but—by exploring its religious, aesthetic, moral, economic and even technological implications—actively contributed to it.

By turning "applied science" into a topic of poetical reflection, Darwin became one of the first literary voices of the "machine age" or "age of industrialization," and, in contrast to William Blake, one that conveyed the optimism initially associated with its potential for the improvement of the human condition. Moreover, he traced the gradual disciplinary dispersion of eighteenth-century "natural philosophy" (singular) into nineteenth-century "sciences" (plural) and their applications in engineering. For Darwin, poetry did not merely provide a medium (or a cultural space) in which to discuss exciting scientific discoveries and developments on the content level. In a far more radical sense, he employed poetry as a genuine scientific research tool in its own right: the poetic imagination became a source for daring scientific speculation and the poetical text an instrument of provocation: Mediating between genres which have only nowadays come to be regarded as irreconcilable, it seriously aimed at stimulating activities in laboratories in order to verify or falsify scientific hypotheses. To dismiss this as nothing but the last gasp of an outdated didactic tradition therefore misses the point and must fall back on the critic who—from a post-romantic point of view—applies an inadequate, ahistorical

---

[85] See Thomson's frequent accentuation of British thinkers, rulers and politicians, poets and scientists (for example in "Summer," ll. 1419–32 and 1479–1579). For Thomson's poetical contribution to British nation building see also Bärbel Czennia, "Nationale and kulturelle Identitätsbildung in Großbritannien 1660–1750. Eine historische Verlaufsbeschreibung" in Ulrike-Christine Sander and Fritz Paul, eds., *Muster and Funktionen kultureller Selbst- und Fremdwahrnehmung. Beiträge zur internationalen Geschichte der sprachlichen und literarischen Emanzipation* (Göttingen: Wallstein, 2000), 355–90, especially 381–6.

measure. Darwin's attempt to write one more didactic poem which is at the same time "useful and entertaining" was more successful than that, and has too often been unjustly ridiculed.

IV: Expansions of the Poetic Atmosphere: Æsthetic and Rhetorical Consequences for Scientific Discourse in the Later Eighteenth-Century Encyclopedia

If with Darwin poetry took its most radical turn towards rhyming science, it is worth noting that an inverse movement can be observed towards the end of the century, suggesting that the scientific encyclopedia also learned from poetry. A closer look at encyclopedia entries, which have so far been quoted mainly in order to clarify the meteorological content of poems, shall further illustrate the cross-fertilization between early scientific and fictional writing during this period. It could be summarized as the embellishment of the scientific discourse with rhetoric gestures, stylistic devices, and narrative strategies traditionally employed in fictional genres like poetry or the adventure story. The entry on "Aerostation" is a good example. Just as Darwin entertains his readers with anecdotes on the triumphs and tragedies of scientific explorers of the air, contemporary encyclopedists liven up their minute descriptions of the latest discoveries in aerology and meteorology by inserting self-contained "embedded tales," whose functional value is by no means factual information. Articles in encyclopedias offer detailed accounts of the fatal accident of aerial navigator Rozier as well as of numerous other sensational and curious events connected with early atmospheric research, revealing how often the course of science took erratic and accidental turns rather than to proceed in a straightforward line of continual "progress."

A report on how adverse weather conditions resulted in the unexpected exploration of the interior of a thunderstorm cloud in a balloon uses deictic elements establishing an illusion of immediacy, formulae suggesting inexpressibility stimulating the reader's imagination, the rhetorical principle of gradual intensification and many other stylistic devices well familiar from adventure stories, in order to create suspense and to convey a sense of awe and wonder:

> The power of ascent with which they set out, seems to have been very great; as, in three minutes after parting with the ground, they were lost in the clouds, and involved in such a dense vapour that they could see neither the sky nor the earth. In this situation they seemed to be attacked by a whirlwind, which, besides turning the balloon three times round from right to left, shocked and beat it so about, that they were rendered incapable of using any of the means proposed for directing their course, and the silk stuff of which the helm had been composed was even torn away. No scene can be conceived more terrible than that in which they were now involved. An immense ocean of shapeless clouds rolled one upon another below them, and seemed to prevent any return to the earth, which still continued invisible, while the agitation of the balloon became greater every moment. In this extremity they cut the cords which held the interior balloon, and of consequence it fell down upon the aperture of the tube that came from the large balloon into the boat, and stopped it up. They were then driven upwards by a gust of wind from below, which carried them to the top of the stormy vapour in which they had been involved. They now saw the sun without a cloud; but the heat of his rays, with the diminished density of the atmosphere, had such an

effect on the inflammable air, that the balloon seemed
every moment ready to burst.[86]

What is more, encyclopedia entries on weather-related termini
increasingly attribute aesthetic qualities to meteorological
phenomena thereby charging them with emotional qualities which
may recall Thomson's enthusiastic reflections on the beauty of
clouds ("Summer," ll. 1371 ff.):

> Thus he was carried up with such velocity, that in
> twenty minutes he was almost 9000 feet high, and
> entirely out of sight of terrestrial objects.... The beauty of
> the prospect which he now enjoyed, however, made
> amends for these inconveniences. At his departure the
> sun was set on the valleys; but the height to which Mr
> Charles was got in the atmosphere, rendered him again
> visible, though only for a short time. He saw, for a few
> seconds vapours rising from the valleys and rivers. The
> clouds seemed to ascend from the earth, and collect one
> upon the other, still preserving their usual form; only

---

[86] Dobson, *Encyclopaedia*, vol. 1, 202, unit 33 (voyage in a balloon by the Duke of Chartres
with two brothers, Charles and Robert, July 15, 1784). A poetic rendering, obviously also
based on an original report by the balloonists, was published a decade before Dobson's
Philadelphia encyclopedia; see poet laureate Henry James Pye's poem, "Aerophorion. A
Poem," in *Poems on Various Subjects*, 2 vols. (London, 1787), vol. 1, 153–61, here 157–8: "Hail
then ye daring few! who proudly soar /.../ To search the extended mansions of the skies.
/.../ Who now his head sublime, astonish'd, shrouds / In the dull gloom of rain-distended
clouds, / And sits enthron'd mid solitude and shade / Which human eye-sight never can
pervade, / Or rides amidst the howling tempest's force / Tracing the volley'd lightning to
it's source; / Or proudly rising o'er the lagging wind / Leaves all the jarring Atmosphere
behind, / And at his feet, while spreading clouds extend, / While thunders bellow, and while
storms descend, / Feels on his head the enlivening sun-beams play, / And drinks in skies
serene the unsullied stream of day." According to a little note, Pye's poem was written on
the occasion of the first ascent of an English aeronaut, Mr. Sadler, in his balloon from the
Physic Garden in Oxford, in November 1784. Many embedded reports of balloonists'
adventures from Dobson's *Encyclopaedia* reoccur nearly verbatim in later published ones, like
Pantologia. A New Encyclopaedia... (London, 1813). For a slightly less dramatic account of the
same and similar events in Howard's *New Royal Encyclopaedia*, see vol. 1, 58–9.

their colour was grey and monotonous for want of sufficient light in the atmosphere.[87]

Further aesthetic evaluations include superlatives like "the most beautiful prospect" (203), "the most enchanting prospect" (203), "a grand and most enchanting view" (204), and "a most captivating appearance" (204). The description of clouds encountered by Mr. Baldwin, ascending from Chester in Mr. Lunardi's balloon on September 8, 1785, employs expanded similes of lyrical dimensions:

> this upper current, he says, was visible to him at the time of his descent, by a lofty sound stratum of clouds flying in a safe direction. The perspective appearance of things to him was very remarkable. The lowest bed of vapour that first appeared as cloud was pure white, in detached fleeces, increasing as they rose: they presently coalesced, and formed, as he expresses it, a sea of cotton, tufting here and there by the action of the air in the undisturbed part of the clouds. The whole became an extended white floor of cloud, the upper surface being smooth and even....Some clouds had motions in flow and various directions, forming an appearance truly stupendous and majestic....Mr. Baldwin also gives a curious description of his tracing the shadow of the balloon over tops of volumes of clouds. At first it was small, in size and shape like an egg; but soon encreased to the magnitude of the sun's disc, still growing larger, and attended with a most captivating appearance of an iris encircling the whole shadow at some distance round it the colours of which

---

[87] Dobson, *Encyclopaedia*, vol. 1, 201, unit 18–19 (evening ascent of Mr. Charles, December 1, 1783). See also the entry on clouds complementing scientific facts with reflections on the beauty of their appearance (vol. 5, 80): "In the evenings after sunset, and mornings before sunrise, we often, observe the clouds tinged with beautiful colours. They are mostly red; sometimes orange, yellow, or purple; more rarely bluish; and seldom or ever green. The reason of this variety of colours, according to Sir Isaac Newton, is the different size of the globules into which the vapours are condensed."

were remarkably brilliant....Again, round the shadow of
the balloon, on the clouds he observed the iris.[88]

A dramatic step-by-step account of Mr. Blanchard's and Dr.
Jeffries's narrow escape from drowning during the first crossing of
the Straits of Dover in a balloon in January 1785 is explicitly
extracted from the minutes of the navigator himself. Instead of a
stylistically neutralized summary of the events which would have
served the purpose of factual information, the encyclopedia offers
a careful selection of particularly exciting snippets from Blanchard's report, using the stylistic device of direct speech graphical-
ly accentuated by inverted commas. Their self-conscious integra-
tion into a text with a scientific claim reveals additional "literary"
ambitions, including the determination to make an emotional
impression on the reader, to emphasize the far-reaching significance of a particular event, to create a sense of authenticity,
immediacy and drama, and to reactivate age-old narrative patterns
like "departure / adventurous journey / supreme test / happy
return," or "struggle against overwhelming odds, overcome by
heroic commitment to a good cause" or more generally, "per aspera
ad astra."[89]

Considering that a later understanding of science would be
closely linked to the idea of specialization and differentiation, and
of the scientific discourse as factual, emotionally neutral, and
objective, what can be observed during the "long eighteenth
century" is a deliberate fusion of the sphere of science and the
sphere of the imagination. Rather than in terms of a clear-cut
separation, the interaction between poetry and meteorology can
be more adequately described as a mutual exchange or give-and-
take relationship. Towards the end of the century this fertile
dialogue increasingly also affected the relations between painting
and meteorology as well as that of painting and poetry, stimulating continual artistic experimentation and innovation. While

---

[88] Dobson, *Encyclopaedia*, vol. 1, 204, unit 35.
[89] Dobson, *Encyclopaedia*, vol. 1, 203, unit 35.

meteorology triggered considerable terminological and conceptual adoptions in poetry, these were reciprocated by successful attempts to poeticize and fictionalize the discourse of the scientific encyclopedia. Following Horace's recommendations in *Ars Poetica*, nonfictional as well as fictional texts on weather successfully contributed to the popularization of scientific ideas by combining the didactic impulse with a claim to good entertainment.

THE MAD SCIENTIST
The Creation of a Literary Stereotype

Barbara M. Benedict
Trinity College of Connecticut

I n a dusty laboratory, shut off from the rest of the world and illuminated by a ghastly light, a lone figure bends over a contraption of vessels, spirit lamps, microscopes, and electrical wiring. Suddenly, he lifts his head fringed with wild, white hair: his eyes gleam fiercely as he raises a bubbling test tube and exclaims to the skies, "Aha! I've done it!"

Who is this figure, where did he come from in Western cultural history, what has he done, and why does he persist in print, film, and legend? He may have discovered the secret of life, or of death, the power of flying, of turning men into animals, of becoming invisible or bionically strong. All of these ambitions characterize the scientific investigator whose devotion to his work has bred a monstrous ambition that has starved away his feelings for his fellow men, for women, God, nature, or even himself. In this paper, I will explore the origin and development of this stereotype and analyze where, how and why he acquired his peculiar traits.

From before Christopher Marlowe's *Dr. Faustus* (1604) to even after Ian Fleming's *Dr. No* (1962), English and American culture has mounted a remarkably consistent representation of the investigator of nature who has gone over the edge. He is conceited, self-absorbed, deeply-schooled yet contemptuous of conformity;

preoccupied by inventions either impractical or destructive; credulous yet learned; alienated from humans and human concerns, particularly sex; male; messily incapable of controlling his own body yet ravenous for absolute power; hungry for immortality. The mad scientist embodies a contradiction: despite his very topic of inquiry, physical nature, he has denied his own physical existence and wants a power beyond nature. In attempting to be more than ordinary man, he has become less. And he instills in others the contradictory feelings of fear and superiority.

The stereotype of the mad scientist grows out of three historical types. First, he is a natural philosopher who explores the universe; he is also an expert on the human body; and finally, he is the incarnation of mental and social defect. By his empirical method of detached observation and repeated experimentation, he embodies the historical challenge to the authority of all social institutions—particularly to religion and politics—that crested in the Renaissance. An independent investigator who takes nothing on faith, his commitment to science poses a threat to the status quo. But his madness predates this method. It denotes the obsessive pursuit of forbidden knowledge, the willfull tossing aside of the good things of this earth and of the promise of the world to come for Satanic knowledge; this is rebellion against the social consensus on what makes a good or proper life. His madness further expresses his refusal to face his own mortality: through conjuring the dead, experimenting with species, and lunar voyages, he seeks escape from the human destiny of death.

The mad scientist was born in the Renaissance from two rebels: the witch and the virtuoso. As he grows up and his parents divorce, he acquires a step-parent: the bumbling physician whose fondness for painful remedies and unnecessary surgery incarnates both his ignorance and a paradoxical hatred of the very human body he has studied. The mad scientist is the professional bent on the destruction of humanity itself.

I. The Madness of Evil:
Scientists Before Science

There were mad scientists even before science was defined. In the Middle Ages before the development of the scientific method of empirical observation and experimentation, men and women who explored the power of nature were considered dealers in the occult. While some acted under the power of the Church, many did not—and even those who did, like Friar Laurence in Shakespeare's *Romeo and Juliet*, whose sleeping potion so mimicked death that it led Romeo to suicide, skirted the edge of the black arts. Throughout England, astrologers, alchemists, cunning men and witches who advised individuals all presented themselves as masters of a special kind of knowledge: the knowledge of nature beyond common nature, the precise kind of knowledge supposedly monopolized by the Church. This included celestial knowledge—understanding the stars and foretelling the future—as well as natural knowledge of the body, sex, birth and death, and medical cures **[see Figure 1]**. Although clergy learned some of these skills, and physicians existed who practiced herbal medicine, village cunning men were the usual source for information about the future or the recovery of lost property.[1]

They were ambiguous figures. Despite pious lip service, these cunning men and women relied on magic mirrors, astrological calculations, and arcane potions to answer clients' questions and ills. Moreover, rather than counseling resignation to the ignorance or punishments that God had given them, they sought to know their fate and recover their losses. Thus, after the Reformation labeled Catholicism superstitious and treacherous, these occult practitioners were considered thieves and villains, a threat to society, religion, even to God **[see Figure 2]**. For example, *The Dreadful Effects of Going to Conjurers* (1710?) reports that when Mrs. Esther Rushway consulted a cunning man about her stolen money

---

[1] Keith Thomas, *Religion and the Decline of Magic* (New York: Scribners, 1971).

and plate, she was doubly punished; after he showed her the thieves in a mirror, the cunning man pocketed the reward she had promised, and informed her that she would never see her money, silver, or the thieves again. And then, when she emerged, defrauded and conscience-stricken at dealing with the devil, God blasted her in a terrible storm as further punishment.[2] By their investigations into forbidden knowledge, these witches usurped the Church's authority, and stirred up people's earthly passions.

All the occult arts implicitly encroach on the Church, some more than others. The alchemist's search for the philosopher's stone conferring immortality ostentatiously ignores the Christian doctrine of resurrection through Christ, while the attempt to turn lead into gold parodies the religious transubstantiation of bread and wine into Christ's body and blood. In confutation of the religious method of gaining immortality through redemption after death, alchemists naturalize miracles, and thus make God the subject of nature. Witches who worship the devil in God's place, astrologers who find fate in the stars rather than in grace, magicians who raise the dead, cunning men, and conjurors were all premodern scientists: they mystified nature yet demystified the supernatural, sought domination by mastery of the physical and metaphysical worlds, and used arcane jargon, rituals, instruments, and materials **[see Figure 3]**. They fostered a popular tradition that dubbed the manipulation of nature immoral, and this tradition was inherited by the satirists of the Restoration virtuosi. Moreover, the threat that occult practitioners posed was also social. They exercised authority over the rest of the village community despite operating outside established institutions, and their professional but mysterious knowledge of the essential aspects of life—health, property, marriage, the future and death—bred fear and respect. The mad scientist inherits this ambiguous social status of illegitimate but compelling authority.

---

[2] *The Dreadful Effects of Going to Conjurers. A Full and True Relation how one Mrs. Esther Rushway...Lost a Considerable Quantity of Plate and Money...* (London, 1710?).

Magical practices with their blend of ritual, dominance, and transgressive inquiry also foreshadowed Restoration science. For example, the best comprehensive guide to magical powers, Ebenezer Sibly's *A New and Complete Illustration of the Occult Sciences*, reprinted from the fifteenth to the end of the eighteenth century, lists skills and spells required not only to recover property and forecast fortunes, but to cure disease and transform substances **[Figure 4]**. The transformation of substances, the forging of value from the valueless, linked magicians, cunning men, and conmen. **[see Figure 5]**.

The key to magical power—as it would be to scientific power—was the ability to explain or, indeed, to overcome death itself. Magicians accomplished this by raising and questioning both evil spirits and benignant ghosts, as the King of Luggnagg does in Part III of Swift's *Gulliver's Travels* (1726). Like digging up corpses for study, harvesting stem cells to clone humans, and dissecting the dead, this taboo act casts the transgressor beyond social intercourse. Moreover, the rituals accompanying this procedure are also significant in showing how the accouterments of mad sciencedom—laboratories, special clothes, jargon, and utter self-confidence—flow from magicians to virtuosi **[see Figure 6]**. After donning his white linen ephod, long black bombazine robe girded by a rune-bestrewn belt, and shoes sewn with crosses, the magician stands in a consecrated circle holding "an holy Bible, printed or written in pure Hebrew," chanting all the names of God. The mystical power here accorded synonyms and foreign words similarly characterizes Witwouds, virtuosi, and quacks from Shadwell's *The Virtuoso* (1676) and Congreve's *The Way of the World* (1700) to Fielding's *Joseph Andrews* (1742). Apparently designed to dazzle the ignorant and thus elevate the expert's status, this jargon eventually signals the loss of touch with humanity. As the virtuoso descends into lunacy, his mastery of technical language overcomes his command of plain English. Like Jerry Lewis in *The Nutty Professor*—or many a professor on a bad day—he becomes unable to communicate.

Because of their knowledge of life beyond death, magicians also resemble mad scientists (and circus majordomos) by assuming absolute authority and demanding absolute obedience. In his attitude toward the subservient spirit, the magician must display authority. He conjures the spirit, *"that without delay of malicious intent, thou do come before me here at the circumference of this consecrated circle, to answer my proposals and desires without any manner of terrible form either of thyself or attendants."*[3] Sibly explains that,

> After these forms of conjuration, and just before appearances are expected, the infernal spirits make strange and frightful noises, howlings, tremblings, flashes, and most dreadful shrieks and yells, as forerunners of their presently becoming visible. Their first appearance is generally in the form of fierce and terrible lions or tygers, vomiting forth fire, and roaring hideously about the circle; all which time the Exorcist must not suffer any tremor or dismay; for in that case they will gain the ascendency, and the consequences may touch his life.   (1104–5)

Calm authority in the face of terrifying sights demonstrates the magician's power not merely over the underworld, but also over his own nature. As Sibly explains, magicians must witness unmoved the spectacles of disaster and mayhem that they have unleashed, knowing they are not real. Inured to horror, the magician sees a reality behind the illusions that ordinary men see. Here lies one cause for the mad scientist's mania for power and indifference to others. Like Dracula, Alexander the Great, or the king, he dwells beyond human pain, and his greed for control overwhelms compassion.

Magicians who penetrate secrets beyond death and make servants of the dead manifest a power unmatched by any other human. Since this power resembles Lucifer's, those who sought it drew the accusation of Satanic ambition. To fearful and religious

---

[3] Ebenezer Sibly, *A New and Complete Illustration of the Occult Sciences* (London, 1790), 1104.

opponents of magic, magicians, by raising the dead, were rulers of the dark realm and thus harbored a desire for power over nature itself. Moreover, these magicians seemed to dabble in death, to make death life, to make a life—make a living—out of death. Their transgression of these fiercely guarded borders between common reality and the unseen, between this world and the next, suggested their transgression of another border: the border between sanity and madness. Like the Renaissance Fool whose madness demonstrates his understanding of a kind of truth denied by the workaday world, the magician himself half belongs in the shadow world of the dead. Similar investigators of death later in history, like scientists and doctors, risk a similar indictment: all are excoriated for insanely preferring another world to this one.

II. From Wand to Telescope:
The Birth of the Bumbling Virtuoso

After the Renaissance, as a scientific methodology started to become systematized, new definitions of both science and madness reshaped the figure of the magician into that of the mad scientist. Where once the Church had monopolized the natural and supernatural, now practitioners of the brand-new method of empirical science did so. Their method ostentatiously discarded inherited knowledge, particularly the categories and theories inherited from Aristotle, along with much astrology, alchemy, and divination. Instead, pioneers of the New Science regularized a democratic method by which to determine the character of natural beings, objects, processes, and forces. This method was close, disinterested observation and sensual perception through taste, touch, smell, sight, and hearing, and repeated experimentation. It made all nature an object of inquiry, and all humans at least potentially inquirers. Although astrologers continued to practice their craft, they increasingly earned the contempt of the educated, like Jonathan Swift whose fictional astrologer Isaac Bickerstaff predicted and triumphantly pronounced the premature death of the successful astrologer Partridge—who was then ludicrously forced to prove in print that he was still alive **[Figure 7]**.

In 1660, at the start of the Restoration of the monarchy following the civil war, a new institution was founded to nurture this methodology, the Society for the Advancement of Learning, the nest in which new-bred scientists were coddled; it was chartered by the King himself two years later, and became an international force in establishing scientific classifications and topics of inquiry. Despite this royal sanction and high visibility, however, the Society remained financially and ideologically separate from the monarchy, unlike similar academies in Europe, acting not as an organ to sound kingly power but rather as the public authority for adjudicating fact from fiction, and answering questions about the secrets of nature.[4] In practice in England there was no central control over the riot of questioning, testing, and observing that the empirical method licensed—nor over the kinds of questions people could ask. Once freed from the moral regulation of the Church, publications and prurient people could peer into any dark corner—into the organs of generation, into the secrets of the universe, into the nature of God. Anyone could do it; almost everyone did. As a result, English culture suffered a crisis of authority over knowledge. If the Church no longer was answering fundamental questions and anyone with eyes could legitimately observe nature or super-nature, who knew anything and how?

The Restoration virtuoso felt he did—or so satirical literature claimed—although he was less clear exactly how. Although Robert Boyle and many others published long treatises and explanations of their chemical and biological experiments, they failed to clarify the scientific method for the popular audience; indeed, since they disagreed on method, meaning, and results, their attempt to find a new language of concrete exactitude mystified not only their readers but apparently themselves.[5] Satirists quickly grafted the

[4] Mario Biagioli, "Etiquette, Interdependence, and Sociability in Seventeenth-Century Science," *Critcal Inquiry* 22 (1996): 193–238.
[5] See for example Boyle's gigantic output in *The Works of Robert Boyle* (London, 1744), particularly *The Skeptical Chymist*; also Sir Kenelm Digby, *Of Bodies, and of Man's Soul. To*

new professor of alchemical transformations and abstruse knowledge onto the old magician with one improvement: whereas most cunning men consciously deceived their clients, the virtuosi unwittingly deceived themselves.

Since they trespassed on God's province, natural philosophers had long been blamed for the traits of arrogance and credulity. Francis Bacon, in his Renaissance riposte to the Medieval distrust both of studiousness and of inquiries into hidden nature, had attempted a refutation. In *The Advancement of Learning* (1605), he formalized the disciplines that differentiated science from magic, classified phenomena into empirical categories, and defined the branches of knowledge and the scientific method. He also categorized the flaws of the scientist—sloth, envy, greed, quarrelsomeness, vanity, and obsessiveness—but Bacon argued that these were either misconceptions of the scientist's activity, or the accidental weaknesses of individual men. The true scientist, he asserted, was virtuous, humble, and pious, dedicated to studying God through nature. Although Bacon's defense offered strong support for scientific endeavors in the following century, the accusations of egoism, malice, and power-mania persisted in the new magician: the virtuoso. One cultural charge, however, was transformed. Whereas Medieval critics had thought natural philosophy bred atheism, Restoration critics believed it bred credulity.

Although they resented the public's neglect, the members of the Royal Society partly understood the skepticism with which they were greeted by both religious authorities and much of general society.[6] Coming after the rancorous civil war in which metaphysical issues were debated with pikes and guns, they were particularly anxious that their empirical method be cleansed of partisan feeling, open, free, and dedicated only to the advancement of information, not to the advancement of any political cause **[Figures 8 and 9]**. Consequently, they passionately asseverated

discover the Immortality of Reasonable Souls. With two Discourses Of the Power of Sympathy, and Of the Vegetation of Plants. (London, 1644).
[6] Michael Hunter, *Science and the Shape of Orthodoxy: Intellectual Change in Late Seventeenth-Century Britain* (Woodbridge: The Boydell Press, 1995), 152.

their disinterestedness: they did not work for money, they did not aim at practical discoveries that would benefit one profession or trade, they did not adhere to a particular philosophy or sect. In order to publicize this purity, Thomas Sprat wrote *A History of the Royal Society* (1667) before it *had* a history in which he proclaimed that the society was constituted of "gentlemen" whose position in life freed them from any particular bias or corruption, and that their joint devotion to truth disallowed quarreling or in-fighting.[7] Unlike magicians, they worked for abstract truths; unlike witches, they sought only good. They were virtuosi: students of art, philosophy, and nature, selflessly devoted to the public weal.

Unfortunately, Sprat's declaration did not fit the facts. Shortly after the establishment of the Society, Henry Oldenburg began publishing a monthly record of scientific reports, named the *Philosophical Transactions* (1660) which encouraged both Members of the Society and like-minded spirits to publicize their discoveries in print.[8] This tremendously influential publication disseminated both the procedures, queries, topics, results, and failures of the New Method, and also the disagreements of the new scientists. A readership avid for gossip and enviously curious about free inquiry into all areas of nature, no matter how sensitive, fell upon any hint of disagreement with delight.

Few disagreements, however, were more public than that concerning the reality of ghosts conducted between Joseph Glanvill, cleric and dedicated Member of the Society, and Henry Stubbe, a fellow cleric. Glanvill, who firmly believed in apparitions, published a long account of ghosts he had verified entitled *Saducismus Triumphatus* (1681) **[Figure 10]**. He even dragooned his fellow Member of the Royal Society, Henry More, into asserting in the preface, "I know by long experience, that nothing rouzes [scoffers and unbelievers] so out of theat dull Lethargy of Atheism

---

[7] Thomas Sprat, *A History of the Royal Society of London, for the Improving of Natural Knowledge* (London, 1667), 67, passim.

[8] Marie Boas Hall, "Oldenburg, the *Philosophical Transactions*, and Technology" in *The Uses of Science in the Age of Newton*, ed. John G. Burke (Berkeley: University of California Press, 1983), 21–47.

and Sadducism, as Narrations of this kind. For they being of a thick and gross spirit, the most subtile and solid deductions of reason does little execution upon them,"yet even they are stung and terrified by these reports.[9] Even though Glanvill's encyclopedia of guaranteed ghosts continued to inform collections of supernatural tales into the nineteenth century, in the seventeenth it seemed to affirm the credulity of virtuosi, who were supposedly yet fallaciously dedicated to "reason," and to show the unreliability of the empirical method and—thanks to More's preface—the snobbish contempt in which they held the rest of the world.

Moreover, the intemperate response of the maddened Stubbe confirmed rumors of the quarrelsomeness and self-interest of virtuosi. In 1670, he published *Campanella Revived: Or, an Enquiry into the History of the Royal Society* alluding to an historic magician, and carrying the subtitle, "Whether the *Virtuosi* there do not pursue the Projects of *Campanella* for the reducing *England* unto Popery." In his "Preface to the Reader" and elsewhere, he accused the Royal Society of desiring to "ruine...the Faculty of Physicians" and apothecaries with new-fangled beliefs designed to aggrandize themselves, labeling them ignorant "Impostours" aiming to "overbear the Universities," and spread atheism or papism.[10] Within months he produced a torrent of other blistering tomes, particularly *Legends no Histories: or, a Specimen of some Animadversions Upon the History of the Royal Society* (1670) ridiculing Glanvill's account of the supernatural, and *The Plus Ultra reduced to a Non Plus* (1670), lambasting the Society's medical methods, anatomy, chemistry, philosophy, experiments, use of telescopes, and lunar voyages, and singling out Glanvill's *Plus Ultra* for ridicule.[11] Telescopes, claimed

---

[9] Henry More, in a letter prefaced to Joseph Glanvil, late Chaplain in Ordinary to His Majesty, and Fellow of the Royal Society, in Joseph Glanvill's *Saducismus Triumphatus: Or, Full and Plain Evidence Concerning Witches and Apparitions* (London, 1681), 12.

[10] [Henry Stubbe], "*Campanella Revived: Or, an Enquiry into the History of the Royal Society,* Whether the *Virtuosi* there do not pursue the Projects of *Campanella* for the reducing *England* unto Popery" (London, 1670), "to The Reader."

[11] [Henry Stubbe], *Legends no Histories: or, a Specimen of some Animadversions Upon the History of the Royal Society* (London, 1670) ; *The Plus Ultra reduced to a Non Plus* (London, 1670).

Stubbe, impair sight when they are not merely useless; they may even delude observers by enlarging common or reflected objects, including the viewer's own eye (*Non Plus*, 26–30). The vaunted, fresh observations upon the moon, he adds, repeat stale knowledge (34–37); in construing its variations of light as valleys and mountains, and further in their "imagination of *Seas* and *Lakes* therein," scientists dream. "'Tis all an *improbable phancie*," states Stubbe, categorically denying the existence of lunar atmosphere, earth, water, colors, or creatures, and subjecting the Society's energetic observations to his own, sour scrutiny (38). Empiricism seemed at war with itself.

Indeed, as Glanvill and Stubbe exchanged fulminating pamphlets accusing each other of ignorance and atheism, their dispute endangered the authority of the empirical method itself. One claimed to have seen, heard, and smelled ghosts, and the other claimed not to have; one school of thought argued on telescopic evidence that the moon had atmosphere, another used the same means to claim it did not. These quarrels defined the new virtuosi as litigious, intolerant, and uncontrolled, helping to lay the foundation for the notion of the scientist as driven by anti-social passions about obscure issues. Like the magician demanding total subservience from rambunctious spirits, the quarrelsome Member of the Royal Society, whatever the theory of communal learning, apparently could not tolerate dissent. Scientists became tarred with the brush of impatient, arrogant megalomania.

The contradiction between this very public quarrel between Henry Stubbe and Joseph Glanvill over the supernatural, and the continuing claim by the members of the Royal Society of impartiality was grist to the satirical mill. Their proclaimed stance of disinterestedness was meant to guarantee the universal utility of their enterprise and discoveries, the solidity of their methodology and research, and the fairness of their approach. In the hands of satirists, however, it assured instead the experiments' pointlessness, and the virtuosi's self-deception, since it meant no experiment need produce anything practical, and no practical test could prove an experiment's worth. The playwright Thomas Shadwell epitomizes this combination in a hilarious scene in his satirical play *The Virtuoso* (1676), which opens with the science-crazed Sir Nicholas Gimcrack, the virtuoso of the title, sprawled on a table, spastically circling his limbs as he imitates a frog in order to learn to swim. He never approaches the real medium of water since, as a "speculative philosopher," his experiments are theoretical: "Virtuosos never find out anything of Use," he proclaims; "'Tis not our Way." Rather, "Knowledge is my ultimate end."[12] In *Gulliver's Travels*, Swift describes a panoply of useless experiments, many of which, like extracting sunshine from cucumbers and original matter from human dung, reverse natural processes for no practical result in a parody of the Royal Society's activities. Gulliver, as a doctor and "Sort of Projector" himself, details these experiments enthusiastically, as mesmerized by science as he is gullible.[13] To satirists, the scientific quest for universal knowledge was a solipsistic dream.[14]

Not only did the virtuosi's posture of disinterestedness seem deluded, but it also suggested that virtuosi were indifferent to the plight of their fellow men, especially if these occupied a dependent social position. In *The Virtuoso*, Sir Nicholas Gimcrack remorselessly dispenses home-made remedies to his tenants and villagers, killing off several with poisonous concoctions, according to his uncle Snarle. He also dabbles in Robert Boyle's new process of blood transfusion by injecting the blood of a sheep into one of his clients, so that the man, now sprouting wool, has become one of Sir Nicholas' "Flock" whom he plans to shear for his own use. At the end, enraged weavers storm the home of the virtuoso, resisting his rumored invention of an automatic loom that will cost them their living.

---

[12] Thomas Shadwell, *The Virtuoso* (London, 1676): V, 84; II, 30. See Marjorie Hope Nicholson and David Stuart Rhodes's introduction to *The Virtuoso*, ed. Nicolson and Rhodes (London: Edward Arnold Ltd., 1966): xxii–xxiv.

[13] Jonathan Swift, *Gulliver's Travels*, ed. Christopher Fox (Boston, New York: St. Martin's Press, 1995), 170.

[14] See Joseph M. Levine, *Dr. Woodward's Shield: History, Science, and Satire in Augustan England* (Berkeley: University of California Press, 1977; rpt. Ithaca: Cornell University Press, 1991), 240–42, passim.

Moreover, because they were dedicated to theory over practice, and because they easily deluded themselves, virtuosi were satirized for sexual inadequacy. Gulliver repels his wife, finding her odor "offensive," preferring a stable to her company, while the frustrated wives of the absent-minded, inefficient, and quarrelsome Projectors in the Academy of Lagado escape to the lush lads below for satisfaction (266, 159). In *Three Hours After Marriage* (1717), a collaborative farce by Gay, Arbuthnot, and Pope satirizing the antiquarian, physician, and member of the Royal Society Dr. Woodward Shields, the virtuoso Fossile marries a whore, and obstinately refuses to believe the evidence of his eyes as she forms liasons with lovers beneath his very nose, and foists a bastard on him as his child: his empiricism vanishes in the face of desires he himself barely recognizes. In "Sir Joseph Banks and the Boiled Fleas," recounting the discovery that fleas are not lobsters, Peter Pindar characterizes Sir Joseph Banks, the President of the Royal Society, as both "A *girl* for novelty"-seeking, and a queen bee, surrounded by effeminate flatterers who service him by returning to the hive with flowers and weeds from around the globe: the members of the Society appear as drones who, "say soft things and flatter, kiss [his] hand...[and] eat the honey for such deeds, and thrive."[15]

The perverse sexuality seen in a passion for abstract knowledge reappears in popular images of the collector, antiquarian, and connoisseur who slavers at paintings of nudes, collects priapic statuettes from Pompeii, and fingers ancient manuscripts. Epitomized by Sir Hans Sloane, who accumulated the gigantic collection that became the British Museum, Richard Payne Knight, author of the *Discourse on the Worship of Priapus* (1786), and the elite Society of Virtuosi who collected ancient art, frequently displaying naked women, caricatures of such virtuosi appear throughout the century. This charge received added power by the coincidence of history. Sir William Hamilton was a renowned collector and connoisseur of especially Roman antiquities, includ-

[15] Peter Pindar, *The Works of Peter Pindar, Esq.* 3 vols. (London, 1805), st.7, l. 41; st. 5, ll. 35–6.

Figure 1: In this frontispiece from the magician Pinetti's *The Conjuror's Repository* (1793), a conjuror reveals the secret of time to a young woman by showing her in a magic mirror how she will look in the future. The hourglass and skull and bones on the table symbolize death while the bubbling cauldron is the emblem of Satanic knowledge.

By permission of the British Library (Cup 406 k 4).

Figure 2: *Humbugging, or Raising the Devil*, aquatint by Rowlandson, 1800; published 12 March 1800 by R. Ackerman. In this satire, a bedazzled client watches a magician conjuring the devil from hell while his hidden, female assistant picks his pocket. Emblems of fraudulent science appear in the preserved alligator hanging from the ceiling, the tortoise and skull in front, and the hieroglyphic stone tablet.
Courtesy of the Print Collection, Lewis Walpole Library, Yale University.

Figure 3: *The Alchemists.* Watercolor by Rowlandson; published 12 March 1800 by R. Ackerman. This satire contrasts natural and unnatural transformation of substance, activity, and generation. While a nubile couple demonstrate natural "Animal Heat" and "Blood Warm" in the back room, two alchemists seek the "Transformation of Metal"; one has a Satanic hooked nose and curling hair that resembles devilish horns. The spirit flasks and suspended fish in the background signal perverse science.
Courtesy of the Print Collection, Lewis Walpole Library, Yale University.

A LABORATORY.
Shewing how a Simple Spirit may be extracted, to represent Flowers & Herbs, in Full Bloom.

Figure 4: *A Laboratory Shewing how a Simple Spirit may be extracted, to represent Flowers & Herbs, in Full Bloom* from E. Sibly's *New and Complete Illustration of the Occult Sciences* (1790), p. 115. This engraving systematizes alchemical transformation into mechanized science, replacing the caricatured, chaotic muddle of early laboratories and curiosity cabinets with a clean, orderly demonstration desk where each item—including the scientist himself—is clearly numbered. By permission of the British Library.

Figure 5: "I. Calculator Esq.: The Celebrated Conjuror" (1776). In Daniel Lysons' *Collectanea* C. 1291. c. 16 (v. 1, f. 2). In this satire on lotteries, an official appears as a magician conjuring devils to forge capital from fraud.
By permission of the British Library.

Figure 6: Frontispiece of Daniel Defoe's *A System of Magic; or, a History of the Black Art* (London: J. Roberts, 1727); 719. H. 14. This satirical engraving portrays a conjuror summoning the devil by inscribing the names of God in a magic circle that protects him; the devil in a jester's outfit enters at the door, indicating the self-delusion of the conjuror. The telescope and globe in the front, bottles and jars over the door, and library books express the fusion of natural science, abstruse learning, and the dark arts.

By permission of the British Library.

Figure 7: *Dr. Silvester Partridge's Predictions*, from *The Works of Mr. Thomas Brown* (London, 1719), vol. 1, p. 163; E. Kirkall sculp. for John Partridge. This advertisement portrays the astrologer surrounded by symbols of knowledge: books, a globe, a staff, stuffed exotic animals, flasks, and charts.
Courtesy of the Print Collection, Lewis Walpole Library, Yale University.

Figure 8: Engraving illustrating Samuel Butler's *Hudibras* (1663). The frontis-
piece, *Sidrophel looking At the Kite Through His Telescope*, satirizes superstitious-
Puritans (inv. Hogarth, sculp. Rowlandson, 1810).
Courtesy of the Print Collection, Lewis Walpole Library, Yale University.

Figure 9: Engraving illustrating Samuel Butler's *Hudibras* (1663). In *Hudibras and Sidrophel,* the frustrated warrior Hudibras punishes his astrologer for false prophesy (etching and engraving, Hogarth).
Courtesy of the Print Collection, Lewis Walpole Library, Yale University.

Figure 10: Frontispiece of Part II of Joseph Glanvill, *Saducismus Triumphatus* (London, 1681). 719.h. 4. This six-part scene depicts four demonic apparitions, solicitations, and levitations, plus visits by the ghost of a deceased father blessing his grandchild and by a guardian angel.
By permission of the British Library.

Figure 11: *A Virtuoso and a Fly*, Woodward del. Cruikshank sculp. (London, 1793). In this engraving, a studious virtuoso feebly "reads" a fly which has flown in from the window onto his book while his stuffed crocodile embodies ferocious nature above him.

Courtesy of the Print Collection, Lewis Walpole Library, Yale University.

Figure 12: Frontispiece of John Wilkins' *Discovery of a New World* (London, 1640). This engraving depicts Corpernicus, Kepler, and Galileo saluting the discovering of "A New World and Another Planet" while the other planets circle the sun overhead.

By permission of the British Library.

Figure 13: *Credulity, Superstition, and Fanaticism: A Medley.* This engraving by
Hogarth depicts a preacher dangling a witch and a devil while a fainting Mary
Toft gives birth to rabbits in the foreground; on the right, a female thermome-
ter charts the progress of the passions from madness to madness.
Courtesy of the Print Collection, Lewis Walpole Library, Yale University.

Figure 14: Willliam Hogarth, *Marriage à La Mode*, Plate 3. In this scene, the rogue Tom pays off a doctor either to cure syphilis or to perform an abortion on the young prostitute beside him. The doctor's coarse joviality reflects that of the grotesque bust on the cupboard; a skull and an alligator indicate the doctor's profession.

Courtesy of the Print Collection, Lewis Walpole Library, Yale University

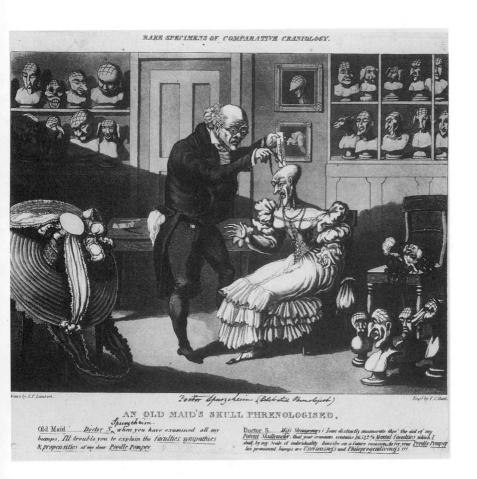

Figure 15: *An Old Maid's Skull Phrenologised,* aquatint drawn by E. F. Lambert, engr. F. C. Hunt, c. 1830. This satire links antiquarianism with medical perversion by showing the celebrated phrenologist Doctor Spurzeheim surrounded by distorted heads, echoed in his own and that of his subject.
Courtesy of the Library of Congress..

Figure 16: *English Credulity, or the Chevalere Morret Taking a French Leave.* L. R. 301.h.3, f. 10. This plate shows a series of frauds that the English have believed, including poltergeists, ghosts, gypsy abduction tales, and Mary Toft, one of the "Rabbit Women."
By permission of the British Library.

ing stones, paintings, and statues excavated from Pompeii where he arrived in 1764.[16] He also possessed a ravishing and sensuous wife, Emma—or rather, did not possess her since she had already had one affair and rapidly became Nelson's lover. Their *menage à trois* helped to confirm the cultural stereotype of the scientific collector as sexually irregular, a man whose sexual inadequacy drove him to find satisfaction in abstruse knowledge. While the mad scientist is not necessarily a collector, he often has the expensive or corrupt tastes of a virtuoso—like the pure-bred white cat purring on the knee of Dr. No and of his parodic twin in Mike Myers' *Austin Powers: Man of Mystery*. Moreover, like the collector whose accumulation of precious objects pleases himself alone, he is greedy, wanting like Faustus fame and power. Sprung from the same historical course, collectors and mad scientists share an unhealthy regard for the rare, unusual, even monstrous.

The Restoration crisis in the authority of knowledge and of practitioners of it vitally informs the emerging trope of the mad scientist. Since, like magicians, the new empirical questioners had no formal training, they lacked a public function in a period that was anxiously defending the significance of the public sphere.[17] Since, too, their methodology could apparently produce contrary conclusions, they did not offer a convincing promise of future good for society. Instead, their quarreling, wealth, and exclusivity made them seem remote from the general weal, secretive, obstinate, and spoiled. Moreover, they patently refused to recognize their own flaws, claiming objectivity while raging over details of data or interpretation, proclaiming disinterest while scrabbling for fame. In the seventeenth and eighteenth centuries, the virtuosi thus failed to forge a practical, social identity, and instead spawned its opposite: the image of the scientist as impractical, self-deluded, and antisocial **[see Figure 11]**.

If the idea of science as the knowledge of nature had changed, so too had the understanding of madness. Before the emergence of

---

[16] Ian Jenkins, "'Contemporary Minds': Sir William Hamilton's Affair with Antiquity," in *Vases and Volcanoes*, ed. Ian Jenkins and Kim Sloan (London: The British Museum Press, 1996), 42.
[17] See Richard Sennet, *The Fall of Public Man* (New York: Alfred A. Knopf, 1977).

science, madness, as Michel Foucault has argued, bordered on divine knowledge because mad people inhabited a sphere which made nonsense of the illusory good things of this world. Such madness was the knowledge of the absurdity, the intolerable contradiction, of a life that must lead to death: knowing the reality of death, the mad person cannot take life seriously. Once the scientific method began to infiltrate medicine in the Restoration, however, doctors interpreted madness as delusion, "error" as Foucault suggests, a physical condition that could be cured rather than an existential perception.[18] For the satirists of the seventeenth and eighteenth centuries, virtuosi bridge these definitions. Like Shakespeare's mad King Lear or mad Fool Tom, they have inquired beyond the realm permitted mankind into the origins of creation. In their exploits into ancient civilizations which mummify the dead, remote countries where alligators roam, deep pits where giants lie buried, and far skies where stars revolve below God, they seek to expose all secrets. Yet, deluded by their scientific obsession, they remain blind to the truths of death and meaninglessness. As Imlac says in Samuel Johnson's *Rasselas* when he surveys the Egyptian pyramids, the creations of the past are "monument[s] of the insufficiency of human enjoyments," testimonials of the pointlessness of existence.[19] Instead, virtuosi believe they see life, worlds to conquer, immortality. Thus, they are victims of gross error, nourished by their ignorance of their own motivations.

Madness as the pursuit of immortality—paradoxically, of death—marks the entire history of the mad scientist from the spirit-raising magician to the sinister chief of the society for the destruction of the world, Ian Fleming's Spectre. The literary figure that epitomizes the tenuousness of the border between impious magic and scientific learning is Doctor Faustus. A scholar versed in divinity, all the natural sciences, poetry, mathematics, philosophy,

18 Michel Foucault, *Madness and Civilization: A History of Insanity in the Age of Reason*, trans. Richard Howard (Random House: New York, 1965), 16–31, passim.
19 Samuel Johnson, *The History of Rasselas, Prince of Abyssinia* (Harmondsworth: Penguin Books, 1976), 108.

and magic, Faustus still cannot answer the human conundrum: what is the meaning of life? In his search for the answer, he sells his soul to the Devil, tastes all the pleasures the earth affords, yet still yearns for more. Faustus's act is moral insanity: at any cost he pursues information, like the ominous captors of Patrick MacGoohan, the Prisoner in the 1960s television series. Faustus exemplifies the madness of willing evil for the sake of unholy knowledge.

While Faustus's madness comes from sin—pride and moral abandonment—even pious figures who seek celestial knowledge also fall into a kind of madness. The quest for heavenly knowledge itself is represented in conventional terms as always mad because, traditionally, knowledge of the heavens is precisely the knowledge denied to man in his time-bound world. Those who seek literally to know the stars miss the true knowledge of them, which is moral: heaven is not the spheres but the angels. In English culture, this ambiguity between physical and metaphysical definitions has contributed to the ridicule of the virtuoso as so rabidly empirical that he cannot see.

The paradoxical blindness of the empiricist informs many seventeenth-century genres that grow into science fiction, in particular the visionary voyage for enlightenment, which originates early in Western history, cresting twice: in the Medieval convention of the dream-vision, and in the seventeenth century lunar voyage. Derived from classical models that flood English culture in the Renaissance, notably Lucian's satiric voyage first translated into English in 1634 by Francis Hickes, dream voyages bring heavenly messages or outlandish experiences; in the seventeenth century, the favorite destination was the moon.[20] As symbols of supernatural ambition and the otherwordly preoccupation of the fractious Civil War generation, and often organized as dream-voyages to circumnavigate reader's disbelief and to link the

---

[20] Classical precedents to moon flight include Platonic myths in Cicero's *Somnium Scipionis* and Plutarch's *De Facie in Orbe Lunare*, and often use the dream device. Five editions of *Satyr Ménippée* (Anon: Paris, 1593–95) appeared in the seventeenth century, with subsequent editions for the next two centuries. See Marjorie Hope Nicolson, *Voyages to the Moon* (New York: Macmillan, 1960), 10–52, esp. 16–20.

celestial with the supernatural world, lunar voyages became a perfect vehicle to satirize the credulity and ambition of virtuosi. Virtuosi's supposed inability to distinguish useful from useless research, near from far sight, telescopic from microscopic worlds lent itself deliciously to the confusion of fantasy and empiricism.

In the mid-seventeenth century, lunar voyages come to represent the virtuoso's delusions of grandeur: the ambition, voracious appetite for knowledge, arrogance, and heedlessness of Faustus, spiced with infantile fantasy. Discoveries through the telescope and progress in hydraulics, physics, and engineering led many members of the Royal Society to consider space travel possible. In lunar literature, writers stimulated by these new discoveries tended to depart from the mystical and allegorical focus of earlier lunar literature in favor of outlining the practical means for reaching the moon. Most notoriously, in 1640 John Wilkins, Bishop of Chester, in his treatise, *A Discourse concerning A New World & Another Planet...A Discourse tending to prove, that 'tis probable there may be another habitable World in the Moone*, speculated on the atmosphere, seas, lands, "high Mountains, deepe vallies, and spacious planes in the body of the Moone" (Contents) **[Figure 12]**. Quoting ancient authorities as Dr. Frankenstein will nearly two hundred years later, he suggests not only that creatures live in the moon, but also in the sun, and that sun inhabitants,

> are like to the nature of that Planet, more cleare and bright, more intellectuall than those in the Moone where they are neerer to the nature of that duller Planet, and those of the earth being more grosse and materiall than either, so that these intellectuall natures in the Sunne, are more forme than matter, those in the earth more matter than forme, and those in the Moone betwixt both. This we may guesse from the fierie influence of the Sunne, the watery and aereous influence of the Moone, as also the materiall heavinesse of the earth. In some such manner likewise is it with the regions of the other starres; for, we conjecture that none of them are without

inhabitants, but that there are so many particular world and parts of this one universe, as there are starres which are innumerable. (190–192)

To putative objections that many sources dismiss these possibilities, and that neither has yet been proven, he unconsciously repeats the magician's jargon, and replies that he desires to "raise up some more active spirit to search after other hidden and unknowne truths." In fact, he argues boldly against following accepted principles: "Since it must needes be a great impediment unto the growth of sciences, for men still to plod on upon beaten principles, as to be afraid of entertaining anything that may seeme to contradict them....Questionlesse, there are many secret truths, which the ancients have passed over, that are yet left to make some of our age famous for their discovery."[21] Such arguments made seventeenth-century scientists seem simultaneously naive and arrogant, even revolutionary, as they pursue their private obsessions by idiosyncratic means.

The echo between the genres of the scientific—or scientistic—lunar voyage and the Medieval dream-vision of heavenly lands highlights a further similarity between dreamers and scientists. Both are driven by ambition, yet both believe themselves humble instruments of a higher good. Following the quarrel between Glanvill and Stubbe in the Restoration, the satirist Samuel Butler wrote a parody of the Royal Society, particularly targeting several key members: Sir Kenelm Digby, the enthusiastic chemist and lunar-watcher, Sir Paul Neale, and John Wilkins, whose *Discourse concerning A New World* evidently caused contemporary hilarity.[22] Entitled *The Elephant in the Moon*, Butler describes the members of the Society sauntering out one "Summer Night, / To search the *Moon* by her own Light."[23] Unaware or

---

[21] [John Wilkins], *A Discourse concerning A New World & Another Planet In 2 Bookes* (London, 1640), epistle to the reader.
[22] Alexander Spence, "Introduction" to Samuel Butler, *Three Poems*, ed. Alexander C. Spence (The Augustan Reprint Society, volume 88; Los Angeles: William Andrews Clark Memorial Library, 1961), iv.
[23] Samuel Butler, *Three Poems; The Elephant in the Moon*, ll. 2–3.

forgetful of the notorious delusiveness of midsummer moonshine, and trembling with excitement, they detect valleys and subterranean passages which they rapidly construe as two areas of inhabitation: while the upper-class lunar beings dwell in the cellars to avoid the sun's scorching rays, "rude Peasants" inhabit the "upper Ground" (ll. 51–2). Thrilled and jealous, another virtuoso snatches the telescope, "Appl[ies] one Eye, and half a Nose" to it, and describes a ferocious civil war raging between the two classes (l. 65). A third seizes the telescope and perceives an elephant, breaking loose from one of the armies and staring nervously back at the virtuosi. As he expatiates excitedly and at length on this discovery as justification for "all our unsuccessful Pains, / And lost Expence of Time and Brains," and suggests publishing a "strange Memoir o'th'Telescope" that will bring them all praise, and other virtuosi debate on the nature of lunar elephants and how they differ from earthly ones, the footboys creep up to peer in the instrument (ll. 177–78, 246):

> When one, whose Turn it was to peep,
> Saw something in the Engine creep;
> And, viewing well, discover'd more,
> Than all the Learn'd had done before.
> Quoth he, a little Thing is slunk
> Into the long star-gazing Trunk;
> And now is gotten down so nigh,
> I have him just against mine Eye (ll. 333–340).

Hastily, one of the virtuosi stares into the tube again:

> He found a *Mouse* was gotten in
> The hollow Tube, and shut between
> The two Glass-windows in Restraint
> Was swell'd into an *Elephant*. (ll. 353–56)

However, the virtuosi have already written their famous narrative about the elephant, at least in their minds. Irritably, they debate the nature of truth, and the finances of their Society, peer again into the telescope, "Review the Tube, the *Mouse*, and *Moon*," and finally resolve to dismantle the telescope and examine it (1. 458). Doing so, they discover hoards of gnats like armies swarming before the lens because "the *Mouse*...by mishap, / Had made the *Telescope* a Trap" (ll. 503–04). Humiliated, they retire.

Butler targets several flaws that will mark the mad scientist. Like Aphra Behn, he portrays virtuosi as deluded by their new-fangled device, the telescope. Indeed, according to the original publisher's note, this incident "was founded upon a Fact mentioned in a Note upon [Butler's famous satiric poem] *Hudibras*...where the Annotator observing, that one Sir *Paul Neale*, a conceited Virtuoso, and Member of the *Royal Society*, was probably characterised under the Person of *Sidophel*, adds—This was the Gentleman who, I am told, made a great Discovery of an Elephant in the Moon, which upon Examination, proved to be no other than a Mouse, which had mistaken its Way and got into his Telescope." (1) Not only are virtuosi bedazzled by new technology, however; they use it to delude themselves.

Moreover, the virtuosi think they pursue truth whereas they really pursue fame. Peering through the telescope with hope-blurred eyes, these virtuosi "take an Invent'ry of all" the Moon's real Estate (5–6). By comparing this survey to "that of *Ireland*," and explaining their purpose to judge whether the land would support plantations, Butler emphasizes the imperialistic and avaricious subtext of their "glorious...Design"(ll. 9, 16). Butler's point is not to derogate an enterprise he and his contemporaries praised for advancing England's wealth and power; rather, he mocks the contrast between the ostensibly selfless purpose of the survey, and the virtuosi's subsequent squabble over who sees the most detail and who gets the most credit. As "all stood ready to fall on" the instrument, "Impatient who should have the Honour, / To plant an Ensign first upon her," Butler further satirizes the unmanly impotence of such an attack against the passive moon (ll. 24–6). This is not heroic battle but sexual transference.

Aphra Behn neatly sums up the range of criticisms against virtuosi in *The Emperor of the Moon* (1687). She portrays a virtuoso Dr. Baliardo peering through a telescope at what he is assured is the lunar emperor who is brooding in his closet for love of the doctor's niece; in a later scene, he is assured of seeing lunar nymphs cavorting naked if he is free himself from sin, and eagerly looks through the telescope again to see them—actually a painting that has been slipped before the lens. He is simultaneously ambitious to dominate the universe by an alliance with the lunar ruler, blind to his own sexual desires, and unable to tell life from art, far from near.

If Doctor Faustus serves as the primary literary precedent for the virtuoso, the pre-eminent historical precedent must be Sir Hans Sloane. A passionate collector not only of curiosities ranging from unicorn horns to Egyptian mummies, but also of other people's collections of curiosities, he wrote dubious medical and chemical treatises, and his cabinet of curiosities held both valuable and conventionally valueless items, which he prized equally. He virtually compelled the nation to accept his collection at huge expense in 1753 when it became the British Museum.[24] Many found the collection absurd: his barber, James Salter, among others, parodied Sloane's greed and credulity with a panegyric:

> Monsters of all sorts are seen,
>   Strange things in nature as they grew so;
> Some relics of the Sheba Queen,
>   And fragments of the famed Bob Crusoe.[25]

Sloane's prominence, first as a member of the Royal Society, then as the founder of a national monument, threw his quirks into relief. His indiscriminate collecting, his idiosyncratic cabinet of the

---

[24] Edward P. Alexander, *Museum Masters: Their Museums and Their Influence* (Nashville: The American Association for State and Local History, 1983), 35; Gavin de Beer, *Sir Hans Sloane and the British Museum* (London: Oxford University Press, 1953), 144–45.
[25] Richard D. Altick, *The Shows of London* (Cambridge, Massachusetts.: Belknap Press, 1978), 17.

rare, the precious, and the dead, and his confusion of the natural and artificial were not unusual at the time; on the contrary, they sprang from a princely Renaissance tradition.[26] Imitated by a private citizen and memorialized in the British Museum, however, they came to epitomize through the following centuries not only England's imperialism, but the virtuoso's mania, overprivilege, and ignorance. Moreover, many of Sloane's items, like those of other virtuosi, were foreign, unconventional, or even traditionally taboo, including strange creatures like the crocodile, natural monstrosities like animals born with two heads, weapons, clothing made of skin, human abortions, and bits of dead creatures. The virtuoso's "cabinet" or room mingled art and nature, life and death, the raw and the cooked. Its rarified collection of disgusting wonders became the laboratory of the madman.

In Restoration satires like those by Shadwell, Butler, and Behn, virtuosi sacrifice exploring human nature—what Pope later termed the "proper study of Mankind"—for explorations into outlandish worlds. In pursuing abstruse knowledge, they overlook the moral and personal knowledge they need. Thus, they remain ignorant of their own human nature: unable to differentiate their hopes from their perceptions, their desire from their sight. Moreover, despite claiming disinterest, they seek conquest and fame, betraying along the way all the flaws of the greedy tyrant: competitiveness, contrariness, obstinacy, credulity, and fantasy.

III. The Professional Scientist: Mad Doctors

As a doctor as well as a natural philosopher and explorer, Gulliver embodies the final element in the creation of the literary stereo-

---

[26] See for example Douglas and Elizabeth Rigby, *Lock, Stock, and Barrel: The Story of Collecting* (Philadelphia: Lippincott, 1944), and *The Origins of Museums: Cabinets of Curiosities in Sixteenth and Seventeenth-Century Europe*, ed. Oliver Impey and Arthur MacGregor (Oxford: Clarendon Press, 1985).

type of the mad scientist. His sanity certainly seems dubious: Christopher Fox has argued that Swift depicts Gulliver himself as suffering from the hallucinations and delusions of melancholia, since his actions like urinating on the burning palace and believing horses speak derive from contemporary literature on madness.[27] But correlative to his own uncertain sanity is his profession, and that profession is key to understanding why he is mad. In the eighteenth century, increasing professionalization of all skills meant that doctors abounded—of divinity, academia, and medicine, as well as natural philosophy—and consequently their claim to special status and abstruse knowledge foundered **[see Figure 13]**.   Albeit all doctors drew satire in this period of anticlericalism and skepticism about science, physicians earned a special animus. Their incompetence, corruption, and cruelty had provided frustration and hilarity for centuries, but the proliferation of medical books and would-be physicians in the Restoration made them a focus for attacks on the failures of science.[28] As early as 1676, the author of *The Character of a Quack-Doctor, or the Abusive Practices of Impudent Illiterate Pretenders to Physick Exposed* excoriated them as "begot in an Illegitimate Copulation betwixt ignorance and impudence."[29] Although doctors had no corporate *professional* identity before the Victorian period, their prominence in culture had been growing from the Georgian period.[30] As Roy and Dorothy Porter explain, "Skepticism—cynicism even—towards doctors was as old as the profession itself. The New Testament told physicians to heal themselves." While the public had difficulty in differentiat-

---

[27] Christopher Fox, "Of Logic and Lychanthropy: Gulliver and the Faculties of the Mind" in *Literature and Medicine during the Eighteenth Century*, ed. Marie Mulvey Roberts and Roy Porter (London: Routledge, 1993), 101–17.

[28] Robert G. Frank. Jr. "The Physician as Virtuoso in Seventeenth-Century England," in Barbara Shapiro and Robert G. Frank, Jr., *English Scientific Virtuosi in the 16th and 17th Centuries* (Los Angeles: William Andrews Clark Memorial Library, 1979), 97.

[29] *The Character of a Quack-Doctor, or the Abusive Practices of Impudent Illiterate Pretenders to Physick Exposed* (London, 1676). 1.

[30] The doctor becomes a staple, if ambiguous, figure in Victorian fiction: Dickens includes dozens of doctors in varying albeit peripheral roles from the innocuous medical student to the actively dangerous fraud in his novels. See David Waldron Smithers, *Dickens's Doctors* (Oxford: Pergamon Press, 1979); also J.C. Da Costa, *Dickens's Doctors* (Philadelphia: Philobiblon Club, 1903).

ing quacks from doctors in Georgian England, they did witness doctors growing sleekly wealthy while they remained in uncertain health [see Figure 14]. The parasitic relationship between doctor and patient—who loses his trade if he succeeds at it—gradually formalized as medics claimed a scientific method and accompanying "professional persona": distant, even chilly, "bustling, prosperous, career-minded," and impervious to patients' opinions.[31] At the same time, key "stars" of medicine captured public attention: Sir Hans Sloane, Richard Mead, John Fothergill, John Coakley Lettsom, the Hunter brothers and so on who made physicians seem very—hazardously—powerful in culture. The "mad doctor" is the mad scientist gone professional.[32]

In the eighteenth century, doctors seemed particularly dangerous for several reasons. One was precisely because their practice had *not* significantly changed, despite the New Science [see Figure 15].[33] Physicians were still associated with magicians and alchemists who aimed to dazzle and deceive, and with cuckolding interlopers when, as midwives, they encroached on male prerogative.[34] By frequently purveying love potions and aphrodisiacs, they stamped medicine with the stigma of illegitimate self-pleasuring. On the other hand, in post-Cromwellian England, medical licensing remained uncertain and fees high.[35]

[31] Dorothy Porter and Roy Porter, *Patient's Progress: Doctors and Doctoring in Eighteenth-Century England* (Oxford: Polity Press, 1989), 54, 55–58, 67–68.

[32] A. J. Engel, *From Clergyman to Don: The Rise of the Academic Profession in Nineteenth-Century Oxford* (Oxford: Clarendon Press, 1983),14, passim.

[33] Roy Porter argues for the progressiveness of medicine in *Doctor of Society: Thomas Beddoes and the Sick Trade in Late-Enlightenment England* (London: Routledge, 1992), but the furor of the debate indicates that there were as many "quacks" as "philosophers"; see esp. 22–36.

[34] Peter Wagner, "The Satire on Doctors in Hogarth's graphic works," in *Literature and Medicine during the Eighteenth Century*, ed. Marie Mulvey Roberts and Roy Porter (London: Routledge, 1993), 200–225. See also Kate Arnold-Forster and Nigel Tallis, *The Bruising Apothecary: Images of Pharmacy and Medicine in Caricature* (London: Pharmaceutical Press, 1989); Ronald Paulson, *Popular and Polite Art in the Age of Hogarth and Fielding* (Notre Dame: University of Notre Dame Press, 1979); Roy Porter "A touch of danger: The Man Midwife as Sexual Predator," in G. S. Rousseau and Roy Porter, eds. *Sexual Underworlds of the Enlightenment* (Manchester: Manchester University Press, 1988), 206–32. Also see the "Introduction" to *Literature and Medicine during the Eighteenth Century*, ed. Marie Mulvey Roberts and Roy Porter (London: Routledge, 1993), 14–15, n. 4.

[35] John Camp, *The Healer's Art: The Doctor through History* (London: Frederick Muller., 1978), 84, 81–82.

Despite doctors' assertions of professionalization, the number of apothecaries and quacks hawking knocked-off Royal Society remedies leapt into the hundreds, multiplying throughout the century, and the demand by angry patients for accessible texts and remedies made the "profession" far from either exclusive or guaranteed.[36] In the 1720s, when Mary Toft bore her brood of seventeen and one-half rabbits and various bits of cat, both her local physician and the king's rushed to her bedside to verify the marvel (one was later convicted of fraud, the other of stupidity). In *Joseph Andrews*, Henry Fielding mocks all profession-als—lawyers, parsons, poets, and actors—but none more than the smug Doctor who pronounces Joseph, who has a slight concussion "a dead Man—the Contusion on his Head has *perforated* the *internal Membrane* of the *Occiput*, and *divellicated* that *radical* small *minute* invisible *Nerve*, which *coheres* to the *Pericranium*; and this was attended with a Fever at first *symptomatick*, then *pneumatick*, and he is at length *grown deleruus*, or delirious, as the Vulgar express it."[37] In *Tom Jones* (1749), Henry Fielding's narrator remarks marvelingly that Tom survives his injuries despite the doctor's best efforts.[38] A key theme in Laurence Sterne's *Tristram Shandy* is the obstetrician's brutal extraction of the infant Tristram from his mother with forceps that injure him in multiple, humiliating ways.[39] As again the Porters have said, "Ever since [James] Gillray" in the late eighteenth century, "leading cartoonists have reveled in portraying society as a sick man being fought over by cantanker-ous and clumsy politicians in the guise of doctors, or rather as

---

[36] Andrew Wear, "The Popularization of Medicine in Early Modern England," in *The Populariza-tion of Medicine, 1650–1850*, ed. Roy Porter (London and New York: Routledge, 1992), 17–41, esp. 22 and 24. See also the huge list of medical practitioners in Peter John Wallis's *Eighteenth Century Medics* (Newcastle-upon-Tyne: Project for Historical Bibliography, 1988).

[37] Henry Fielding, *Joseph Andrews, with Shamela and Related Writings*, ed. Homer Goldberg (New York,: W. W. Norton, 1987), 50.

[38] Henry Fielding, *Tom Jones* 8.13

[39] For an account of a feminist eighteenth-century midwife, see Isobel Grundy, "Sarah Stone: Enlightenment Midwife," in *Medicine in the Enlightenment*, ed. Roy Porter (Amsterdam : Rodopi Press, 1995), 128–144.

'state quacks'."[40] Doctors earned distrust for their mystification, ignorance, prurience, and charlatanry **[see Figure 16]**.

Moreover, in the early modern period, this new figure becomes a literary trope: a feature of all kinds of writing from drama, to satire, to poetry, into the novel. This is partly because the topics he addresses dominate enlightenment prints—from the *Philosophical Transactions* to *The Athenian Mercury* to pornography to manuals on how to fill hot-air balloons. These topics dominate literature, indeed, as a consequence of the lack of disciplinary boundaries in the period: scientific writing was *part* of literature, and the mode for scientific communication was publication. Indeed, the notion that the author resembles a doctor character-ized one aspect of author-reader relations.[41] Partly, literature's hospitality to the mad scientist figure results from the historical overlaps between the new language and categories of science and that of folklore, magic, and gossip—particularly concerning the body, sexuality, hysteria, death.[42] Springing out of the Renais-sance's propulsion toward new enterprises and kinds of knowl-edge, this figure represents a new kind of hero for an increasingly private, print-perusing culture. Not a soldier or statesman, he seeks knowledge for its own sake—or, rather, for his own pleasure.[43] Privileged by wealth, leisure, and social connections, the virtuoso pursues whatever interests him, an amateur who defines impor-tance according to his own whim, often a prince or nobleman whose collections of wonders served to represent the world in

[40] Dorothy Porter and Roy Porter, "Introduction" to *Doctors, Politics and Society: Historical Essays* (Amsterdam: Rodopi Press, 1993), 1; see also Janet Semple, "Bentham's Utilitarianism and the Provision of Medical Care, "42 and passim (30–45).
[41] Marie Mulvey Roberts and Roy Porter, "Introduction" to *Literature and Medicine during the Eighteenth Century*, ed. Marie Mulvey Roberts and Roy Porter (London: Routledge, 1993), 1–13.
[42] Johanna Geyer-Kordesch, "Whose Enlightenment? Medicine, Witchcraft, Melancholia and Pathology" in *Medicine in the Enlightenment*, ed. Roy Porter (Amsterdam and Atlanta, GA: Rodopi Press, 1995), 114. Geyer-Kordesch observes that physicians used folktales and superstitions in their diagnoses (113–27).
[43] Walter E. Houghton, Jr. "The English Virtuoso in the Seventeenth Century," in *Journal of the History of Ideas* III, Part I (January-October, 1942), 51–73 and Part II (April, 1942), 190–219, esp. Part I, 54–65.

miniature.[44] English virtuosi, coming late to the game, characteristically accumulate objects in private collections, and study rather than socialize, to avoid melancholy rather than to model social power.[45] Incarnating the very essence of high culture, however, the early seventeenth-century virtuoso embodies the learning, progress, and power of the Renaissance. Although magicians and witches might be of either sex, the virtuoso is increasingly a male.

However, the defective doctor does differ from the mad scientist. Peter Wagner has described the codes that characterize doctors in eighteenth-century satire, particularly their obsession with their own reputations, cruelty, and habit of making off with the bodies of those hanged at Tyburn. This practice induced riots and resulting legislation decreeing that judges could order dissection as part of the criminal's punishment, which Hogarth depicts as still more savage in his engraving *The Four Stages of Cruelty*.[46] Whereas doctors defraud their clients by practicing a skill they know is fraudulent, scientists believe in their fantasies: they are clients of their own bag of tricks. Rather than cheats, they are self-deceived: they begin as Dr. Jeckyll and end as Mr. Hyde. They embody the professionalized soul: the soul made mad.

In post-revolutionary English culture, Mary Shelley created a fresh archetype of the mad doctor: Victor Frankenstein. Frankenstein is the virtuoso seen through Romantic spectacles: ambitious, obsessed, self-absorbed, he pursues discredited, scientific "authorities" like the magicians Paracelsus, Albertus Magnus, and Cornelius Agrippa (albeit he discards them), and disregards his health, family, friends, and own instincts to make an abortion of nature, to reverse death and life. Crosbie Smith argues that "Frankenstein's obsession, his isolation, his individualism, and his egoism are strongly suggestive of Romantic images of the mad genius, the

[44] For an analysis of the flaws of the virtuoso, see Barbara M. Benedict, *Curiosity: A Cultural History of Early Modern Inquiry* (Chicago: University of Chicago Press, 2001), esp. chapter 1.
[45] Houghton, I, 64.
[46] Peter Wagner, "The Satire on Doctors in Hogarth's graphic works," in *Literature and Medicine during the Eighteenth Century*, ed. Marie Mulvey Roberts and Roy Porter, 207, 212–13. Dorothy and Roy Porter, *Patient's Progress: Doctors and Doctoring in Eighteenth-Century England* (Cambridge: Cambridge University Press, 1989), 54–65.

creative artist, and the natural philosopher *qua* natural magician."[47] Albeit Shelley consulted contemporary science, particularly electricity, when discussing how the doctor made the monster, and departed from prior models to create something that was "the product of a scientific experiment, not a religious ritual," Frankenstein embodies what magicians become in the age of unbelief: students so deeply learned in nature that they transcend or transgress it.[48] Upon his failure to raise ghosts, he rejects alchemy—but his desire to revive the dead remains. Both this desire and its solution also remain secret, like alchemy itself and other methods of attaining forbidden knowledge.[49] Moreover, by transforming flesh into being, Frankenstein does alchemy or magic, conjuring the imitation of life, a body without a soul which becomes, horribly, a mad soul within a grotesque corpse. Significantly, Frankenstein is often confused with his creation: the mad scientist becomes indistinguishable from his mad creation. Both are part of the same violation of nature.

IV: Conclusion

Why has the figure of the mad scientist stayed so gripping, so central to our fears about our culture? That he is central seems undeniable. Mad scientist movies abound from *The Cabinet of Dr. Caligari* to *The Return of the Mummy*; a Hollywood collection of famous film shots made into postcards is entitled *Mad Doctors, Monsters and Mummies!*[50] Versions of Frankenstein reduplicate,

---

[47] Crosbie Smith, in *Making Monstrous: Frankenstein, Criticism, Theory*, ed. Fred Botting (Manchester and New York: Manchester University Press, 1991), 41.
[48] Samuel Holmes Vasbinder, *Scientific Attitudes in Mary Shelley's Frankenstein* (Ann Arbor, Michigan: UMI Research Press, 1984), 50, 51–63, 67. See also Anne K. Mellor, *Mary Shelley: Her Life, Her Fiction, her Monsters* (London: Routledge, 1988), esp. 89–114; also Crosbie Smith, "Frankenstein and Natural Magic" in *Frankenstein: Creation and Monstrosity*, ed. Stephen Bann (London: Reaktion Books, 1994), 39–59.
[49] Markman Ellis, *The History of Gothic Fiction* (Edinburgh: Edinburgh University Press, 2000), 149–53.
[50] Seymour Simon and Denis Gifford, *Mad Scientists, Weird Doctors, & Time Travelers in Movies, TV, & Books* (New York: J. B. Lippincott, 1981); *Mad Doctors, Monsters and Mummies!*

spawning dolls (including one that drops its underpants when it blushes), mystery games, masks, and myths.[51] The *Mad Scientist: Riddles, Jokes, Fun* game book uses the trope to sell sick puns. The cliche is so stale we scarcely see it any longer. Why?

Many answers exist. Frank Botting has suggested that "Popular Frankenstein fictions reinforce the technological anxieties attributed to the novel: unfortunate humanity is besieged by the uncontrollable technological monster created by an irresponsible science."[52] Another, related answer has to do with second category itself: the scientist. The activity of "science" moves in history from the periphery to center stage. It begins as an amateur vocation—itself in many ways threatening to more conventional models of elite leisure activity—to a fiercely competitive, strictly monitored and regulated, highly paid profession in a period from the 1780s to the present when professionalism itself was transforming, perhaps dehumanizing, social relations: professionals whose responsibility always lies elsewhere, who are always following orders, incarnate the danger of process without moral supervision over product. The scientist as a professional is thus allied to amorphous, indeterminable interests beyond himself, which allow him to carry out his individual perversions. One sociologist suggests, "The twentieth century has become the century not of the common man but of the professional expert."[53]

Why are scientists particularly prone to insanity in the representations of Western culture? One reason is the illegibility, construed as secrecy, surrounding science. Despite modern attempts at openness, scientific information—on how to avoid contracting diseases like AIDS, on anthrax, cloning, mapping the human genome, and on chemical resistance to disease in

(London: H. C. Blossom, 1991).

[51] Radu Florescu, *In Search of Frankenstein: Exploring the Myths Behind Mary Shelley's Monster* (Robson Books, 1975, 1996), xii, passim. Interestingly, he traces one of the "first authentic Frankenstein alchemy tales" concerning unearthing a dead body in 1763 (87–8).

[52] Fred Botting, "*Frankenstein* and the Art of Science" in *Making Monstrous:* Frankenstein, *Criticism, Theory,* 164.

[53] Harold Perkin, *Professionalism, Property, and English Society Since 1880* (Reading: University of Reading, 1981), 23.

cattle—rarely translates clearly into mass culture. The procedures involve both a knowledge unavailable to most people, and an indifference to blood, guts, and death traditionally deplored. Science conceals its means and reveals its results, but the tradition of hidden experiments and failed "cures" casts around these results a web of mystery. Just as magicians and virtuosi vaunt the danger and privilege of their special knowledge, the scientist (or doctor) with his own version of magical language typically cannot explain his information to the uninitiated. When he does so, it seems to many that beauty collapses: consider Humbert Humbert's scientific description of a dimple in Vladimir Nabokov's *Lolita* as the adherence of the outer layer of tissue to the inner layer. Represented in popular culture, the scientist, insulated by his own empiricism, is cut off from a world that celebrates time he lives in a world of knowledge below the skin; he negates beauty; he is Doctor No. Such a position invites, or resembles, madness—the rejection of our world for a world of one's own.

In the machine-dependent twentieth century, the mad scientist encapsulates the culture's resistance to this kind of dissection of traditional values. Highlighted by the gruesome events of history—Nazi "doctors" experimenting on the inno- cent—scientists and doctors have come to symbolize the heedless pursuit of some self-defined idea of "progress" at the expense of civilization and humanity. The Doctor in Stanley Kubrick's *Strangelove: Or How I Learned to Stop Worrying and Love the Bomb* models the death-loving pervert, and television in the 1950s to the 1970s, when America and Russia seemed to be speeding toward nuclear war, abounds in related images: the Incredible Hulk, a chartreuse muscleman hiding within the benevolent, hippie Doctor; the frigidly inhuman Dr. Spock in *Star Trek*; the neurotic plotter Dr. Smith in *Lost in Space*. Even in the 1980s and 1990s, the stereotypical, humoral imbalance of scientists feeds popular culture. Stephen Spielberg brilliantly captures this cultural ambivalence in his cartoon *Pinky and the Brain*, a half-hour, animated drama about two over-bred white lab rats, one of whom has degenerated into febrility and endlessly spins, grinning, on his hamster-wheel, while the other, short and stocky with a huge, pulsing brain, grips his cage bars, and stares grimly out. At the start of every episode, as night falls, the dippy Pinky asks brightly, "What are we going to do tonight, Brain?" His hunched companion replies thrillingly in cultivated, deep accents, "What we do *every* night, Pinky: **try to take over the World!!**"

The mad scientist embodies science-engrossed Western culture's responses to the desire to change nature. Wand-wielding to microscope-managing to test-tube tasting, the mad scientist represents the danger of mastering forbidden knowledge.[54] In moments of cultural anxiety, the quest to penetrate the unknown and fiddle with the order of things denotes a perverted ambition to become more than you should be, to control or corrupt nature as you are yourself corrupted. Here stands the arrogant Restoration virtuoso who transfuses dog blood into humans and his heirs, Dr. Moreau who forges humans from beasts, and Dr. Jekyll who makes himself a beast. Here, too, stand doctors whose power over death and life we have deified and demonized, in fiction and in fact. For Dr. Frankenstein and Dr. Strangelove, there is Dr. Kavorkian, and the doctor who carved his initials in the belly of his patient after her caesarian delivery.

As Oscar Wilde noted, life imitates art; in history as in literature, the preoccupation with physicality that marks the scientist and doctor invites the cultural charge of perversion. Popular culture has also fetishized physicians turned murderers. In the sexually repressed and obsessed late nineteenth century, the press delighted in deadly doctors: Dr. Crippen the bloody; the pathologically jealous Dr. Buck Ruxton or rather Bukhtyar Ruttomji Ratanji Hakim who dismembered his wife Isabella and maid Mary Rogerson—"I was Mad at the time," he said—and bundled the heads, arms, fingers and so on together neatly in parcels; the philandering, blackmailing Dr. Thomas Neill Cream, who poisoned prostitutes with a strychnine aphrodisiac in South

---

[54] Roger Shattuck, *Forbidden Knowledge: From Prometheus to Pornography* (New York: St. Martin's Press, 1996).

London; the sanctimonious Dr. Edward William Pritchard, who murdered his wife and mother-in-law with an agonizing potion and watched them die with love on his lips; Dr. George Henry Lamson, a crippled drug-addict whose flushed face and staggering steps appeared madness, but who cooly fed his eighteen-year-old brother-in-law poisoned cake for a paltry inheritance; the puny, lecherous Dr. Robert Buchanan, who poisoned his wife and went out on a pub crawl while she died.[55] All of these doctors transgress social and natural borders, and therefore they are all called mad. Their madness is the divine knowledge of death, and also the desire to taste it, deemed divine in martyrs but satanic in warriors. This desire inspires the most powerful, modern images of the agency of death in film and fact: Slim Pickens riding a nuclear warhead to the earth in a patriotic frenzy; Atta boarding the airplane bomb and the twin towers crumbling. It is thus madness for us to ignore them. For physicians and scientists have come to symbolize society's health and illness: our future. In the twenty-first century, the trope will continue as we face—if not worse—replication by clones raised in petrie dishes from cannibalized cells, and consume genetically modified Frankenfoods. But let us hope also for other changes: perhaps the next generation of ambitious scientists will be mad for life, not death, and perhaps in literature that madness will appear in a new trope.

In a wide, light room with painted, colorful walls, a group of scientists are talking. Suddenly, two of them turn to each other and fling up their hands, laughing: "Aha! We've done it!" they exclaim as the woman leans over to kiss her friend. What has she done? Perhaps she has saved the world.

[55] Max Marquis, *Deadly Doctors* (Great Britain: Warner Books, 1992), 78, 5; 92–3; 127; 185, 190, 224; 227, 242.

ELASTIC EMPIRICISM,
INTERPLANETARY EXCURSIONS,
AND THE
DESCRIPTION OF THE UNSEEN
Henry More's Cosmos,
John Hutton's Caves,
and
Georg Friedrich Meier's Quips

Kevin L. Cope
Louisiana State University

T he empiricist movement of the seventeenth and eighteenth
centuries shows a surprising lack of enthusiasm for the
"experience" that was allegedly the focus of its attention. Not
that "empiricists" are lacking in sensuous delights: there is more
than enough to see and enjoy among the diverse texts and wide-
ranging speculations of the advocates of sense, even if most of
what is offered stands at one or more removes from immediate
experience. The core texts of empiricism, whether those of Locke,
Berkeley, Hume, or any of the continental philosophers between
Descartes and Kant, unveil many varieties of imaginative literary
discourse, whether essays or reflections or rhapsodies or dialogues

or discourses. What we don't find in equal abundance are matrices of experimental evidence or heaps of technical reports or even much in the way of experientialist writing other than occasional travel accounts or anecdotes ascribed to assorted authorities. True, John Locke reports on exotic birds, Robert Hooke escorts his audience into the miniature world of seashells, and Adam Smith records market prices, yet experiential reportage accounts for only a minuscule part of the empiricist canon.

If we take the term "empiricism" in its more colloquial, more literary meaning, as an experientialist mentality influencing many aspects of seventeenth- and eighteenth-century culture, and if we apply that enlarged meaning to the huge cache of works produced under the influence of "the new science," we still find more talk than data, more art than experience. Relentlessly descriptive poems like James Thomson's *The Seasons* spend many a line explaining *how* experience *should* be described; sensationalist works such as Robert Blair's *The Grave* take one glance at experience and then recoil into descriptions of our horrified reaction to it; still other putatively experientialist works such as Shaftesbury's *Characteristicks* cultivate an "enthusiastic" rhetoric better suited to thespians than to field researchers or laboratory technicians. In these and many other cases, the description of techniques for describing experience overwhelms experience itself.

This slate of counterexamples to the assumption that empiricism and experience go together is by no means exhaustive or systematic, but it does suggest that empiricism had a harder time finding and promoting experience than its propagandists advertise. The empiricists and the various writers in their milieu may have aspired to replace abstract scholastical philosophy with fresh "experience," but in practice they were rather more eager to investigate the grounds, the counterparts, the surroundings, the contexts, the presentations, and even the negations of the sensible world. By developing theories and definitions as to what was tangible and what was mere chimera, those in the empiricist school inadvertently drew attention to what was *not* part of the sensory field. For John Locke and Immanuel Kant, for example, short paragraphs defining and explaining sensually intuited ideas are dwarfed by extended passages on intellection, reflection, language, communication, and a host of other semi- and non-empirical matters. Many of the extra-experiential interests of the empiricists veer so far from "sense" as to border on embarrassing. The Royal Society's interest in witchcraft, an interest that many scholars find puzzling in a cadre of self-styled enlightened scientific virtuosi, is one such tangent of this "empiricist" exploration of the perimeters, backgrounds, frames, and contexts for the visible world.

As I have explained elsewhere,[1] empiricism had trouble determining what was meant by key words such as "idea," "sense," or even "experience." For Locke, defining the word "idea" required a three-part taxonomy of different types and levels of "idea" along with assorted special instances, waivers, qualifications, and sub-definitions. Like his precursor Sir Francis Bacon, Locke was lexicographically surrounded by older, less austere usages of the word "idea," usages that resonated of Plato, Ficino, Italian Humanism, and, within the English literary tradition, of Spenser and his neo-chivalric allegories. Locke tried hard to straddle the boundary between "ideas" in their limited, modern, technical sense and much bigger "ideas" in the grand old Socratic sense; as a result, he was driven to declare big and complex concepts like "liberty" or "drunkenness" to be just as much "ideas" as more modest notions or perceptions like "red" or "hardness." Locke allowed terms like "ideas" or "experience" to cover a huge size range and conceptual spectrum, one running from localized sensory input all the way up to highly refined abstractions. Locke's is by no means the only case in the international story of empiricism, but it highlights the technical difficulties experienced by the empiricist movement in figuring out exactly what "experience" might be and foreshadow-

---

[1] See "Locke, Mandeville, and the Insignia of the Future: Terminators, Mutants, Vectors, Plurals, Emblems, Maps, Targets, Proposals, Narratives, Crawfish," in Carla Hay and Syndy Conger, eds., *The Past as Prologue: Essays to Celebrate the Twenty-Fifth Anniversary of ASECS* (New York: AMS Press, 1995), 245–80.

ing the kind of elaborate, appealing imprecision that could occur in the less technical discourse of empiricist-inspired literature.

To add to the paradox of nonempirical empiricism, one major goal of this movement was to deal with things that had not yet been or that could not be detected. Procedurally if not methodologically, early empiricism was split between the refocusing of attention on day-to-day, easily accessible experience and probing into unknown, presently unseen places and spaces. One quaint example of this concern to reach beyond available experience is the suggestion voiced by both Abraham Cowley and Bishop Sprat, in their respective *Proposition for the Advancement of Learning* and *History of the Royal Society*, that educational institutions should maintain a fleet of ships to carry professors to places where they might collect previously unknown sorts of data (a plan that modern granting agencies might want to consider!). This impulse to reach out for unknown experience fostered numerous speculative accounts of potential journeys to the moon or the planets, whether John Wilkins's *Discovery of a World in the Moone* or Gabriel Daniel's *Voyage to the World of Cartesius* or Samuel Butler's whimsical *An Elephant in the Moon*; it sustains John Dryden's suggestion, in *Annus Mirabilis*, that the English "make discoveries where they see no sun" and that an English ship, sailing over the "ocean leaning on the sky," might eventually land at an interplanetary freeport and "on the lunar world securely pry."[2] Nor was great or visible magnitude needed to enter this highly exploratory sort of "experience"; poets like Lady Margaret Cavendish searched for worlds within sub-atomic space, inside the molecular structure of rings or the atoms in locks of hair.[3]

Whether in verse or vituperation, essay or effusion, the extraexperiential elements in empiricism promoted an interest in *invisibilities*, in "experiences" that are either not visible or that would remain invisible without heroical scientific efforts. Visible experience, after all, always implied some degree of invisibility. By

[2] *Annus Mirabilis*, lines 640, 654, 656.
[3] See, for example, Margaret Cavendish, *Poems and Fancies: 1653* (Menston: The Scolar Press, 1972), 44–46, "Of many Worlds in this World" and "A World in an Eare-Ring."

directing attention to the way that things actually looked ("were perceived") rather than to (neo-Aristotelian) underlying essences, the empiricists found that visible objects had a nasty way of blocking our view of other objects. Looking at a teacup meant occluding part of its saucer; training a telescope on Jupiter meant blocking the view of the star behind it. To deal with these obscuring tendencies in experience, the empiricists began inventing their own set of para-empirical properties, including such obstructive attributes as solidity, opacity, and hardness. All of these properties were to be discovered inferentially rather than sensually. The inability to see through a steel girder, for example, indirectly proved that it had the property of opacity. All the properties in this category emphasized the invisible and the negative—for example, the fact that one could *not* push one's hand through a brick wall or could *not* walk through it to the other side implied that the wall had "solidity." The empiricists took a keen interest in the philosophical problem of the identity of persons or objects partly because the recognition of an identity requires setting out perimeters for any perception. To know what an object both is and is *not*, there must be some point or line or tangent or boundary at which one object stops being visible and another comes into view. This preoccupation with the identity problem is but one instance of the preeminence in empiricism of perception, cognition, and what John Locke calls "understanding"—a cognitive and epistemological emphasis that entails enriching the "empirical" with an array of faculties, entities, substances, beings, powers, or creations that promote sensory experience but that never appear within it, that subsist in a non-sensible world above, beneath, around, or otherwise outside the visible one.

Presented as a preliminary exploration, this paper will study a small selection of Enlightenment thinkers and writers with invisibilitarian inclinations: Henry More, the later seventeenth-century speculatist, Cambridge Platonist, pop scientist, and witch hunter; John Hutton, the inveterate later eighteenth-century visitor of caves and other dark places; and Georg Friedrich Meier, a mid-eighteenth-century German expert in scientific theories of imprecise or elusive phenomena, his most notorious topic having been the æsthetics of jesting. Although far from a complete or representative sample, these three figures, all of them highly popular in their own times, open a window on the sometimes literary, sometimes philosophical, and occasionally artistic examination of the edges, fringes, and invisibilities of empiricism. These writers peer into aspects of experience that elude the usual tools of empiricism, that define a new science of excess and sloppiness, of things that cannot be fully seen, measured, or analyzed. For More, this "elastic" dimension of empiricism, this ability to stretch around the visible into the rim of the invisible, becomes a cosmological system; for Hutton, it becomes a ground for romantic enthusiasms; for Meier, it serves as a vehicle to test the limits of science itself by setting an ironic, counter-scientific, undisciplined phenomenon like laughter at the very center of philosophical disciplines. These test cases sketch out the horizon of a dark poetics and twilight philosophy on the other, invisible side of experience, a side to which phenomenalism was to give a scientific form in the nineteenth century and that Jane Austen and and Matthew Lewis were to spoof in novels like *Northanger Abbey* or *The Monk*.

I: Henry More's "Spacious Hall" and the Productivity of the Deep Cosmos

At first glance, Henry More seems the least likely candidate for an inquiry into invisibility.[4] If any charge can be brought against

---

[4] Indeed, it seems as though Henry More is an unlikely candidate for any variety of academic study, in large measure because his writings are too densely philosophical for most literary scholars and too extravagant and eccentric for professional philosophers. In his eccentric way, More is highly representative of the interdisciplinary culture of the later seventeenth century and of the resistance of major writers such as John Bunyan, the Cambridge Platonists, or even John Locke to assimilation into twentieth- and twenty-first century academic departments and intellectual compartments. The small body of academic and

More, it is certainly not that of thinness, scantness, or emptiness. More believed in more. Both ardent experimentalist and Platonist idealist, he sought to explain everything from the origin of human souls and the genesis of the universe to the rules for writing love poetry. The industrious More added to his challenges by doing all of the above in seemingly the least eligible of genres, the Spenserian stanza. Spenser's epic measure, More thought, would enhance his cosmological divagations with an air of romance and a patina of antiquity.[5] Cramming troves of knowledge and caches of speculation into Spenser's tightly interweaving rhymes, More leaves his readers with the impression of busyness, satiety, and congestion. If all this were not enough, More, whose overfilled research schedule once led him to remain in uninterrupted cogitation for an entire year in his Oxford study room, appends to his philosophical epic, *A Platonick Song of the Soul*, a supplemental poem, *Democritus Platonissans, Or The Infinitie of Worlds*, in which he announces that he has really only covered one topic, that an infinity of additional worlds remains available for description. With enough lines of argument to dizzy a Jackson Pollack, More hardly needs invisible realms to fill out his daily toils.

It may be that More's attainment of the philosophical and poetical equivalent of gridlock is responsible for his probe into openness and unseen spaces. In order to find room for still more

---

largely biographical writing on More includes the "Introduction" to Flora Isabel Mackinnon's *The Philosophical Writings of Henry More* (New York: Oxford University Press, 1925); the introduction to Geoffrey Bullough's *Philosophical Poems of Henry More* (Manchester: Manchester University Press, 1931); and the extensive biographical, philosophical, literary, and textual analysis offered in the preliminary material to Alexander Jacob's edition of More's *A Platonick Song of the Soul* (Lewisburg: Bucknell University Press, 1998). In 1990, a collection of essays, commemorating the 300th anniversary of More's death, appeared (Sarah Hutton, ed., *Henry More [1614–1687]: Tercentenary Studies* (Dordrecht: Kluwer Academic Publishers, 1990). This volume contains several useful essays, albeit mostly essays on specific scientific and philosophical controversies within More's very large œuvre.

[5] More was not altogether singular in recruiting Spenser and Spenserianism into unusual projects. Spenser imitation was a widespread, varied, and underappreciated phenomenon throughout the seventeenth and eighteenth centuries. For more on More's and other's use of Spenserian diction and meter for offbeat purposes, see David Radcliffe, *Forms of Reflection* (Baltimore: Johns Hopkins University Press, 1993).

topics, he needs to find some way out of his post-Spenserian jam-up. In a very literal way, the overflowing More needs more empty space in which to decompress the surplus ideas that he records from all too many sources. It is difficult to pin down exactly what More believes because he boasts about his collegial ability to change his mind. We can, at least, take More's word for his hope to "Platonicize" "Democritus" by expounding an idealist account of the origin of the material world, by reconciling Hermetic mythology with seventeenth-century atomism. In *The Platonick Song of the Soul*, More elucidates the history of the world from its psychological and generative side. He explains how the creative effluvia of the Prime Mover engendered an array of post-Plotinian, crypto-Christian emanations such as Æon, Psyche, Physis, and Hyle, along with a few sundry planets, moons, comets, and meteors. In his follow-up piece, *Democritus Platonissans*, More takes a more Aristotelian approach, looking *post-hoc* at how the world must operate given information gathered to date about the aforementioned entities. Eschatology is never far away as More probes the life cycles of worlds within the infinite universe. *Democritus Platonissans* thus develops a kind of empirical science based on neo-Platonic assumptions, extrapolating from direct observation to conclusions about the grounds of experience and about possible experiences beyond our present ones.

Setting the tone for his plunge into the unseen, More opens his philosophical romance with a curious non-image of non-sound within a trumpet:

> in silver trump I sound
> Their [the objects in the infinite universe] guise, their shape,
>   their gesture and array,...
>   as in silver trumpet nought is found
> When once the piercing sound is past away
> (Though while the mighty blast therein did stay,
> Its tearing noise so terribly shrill,

That it the heavens did shake, and earth dismay).[6]

More's first purpose in *Democritus Platonissans* is to explain his about-face regarding the infinity of worlds, a thesis that he had previously disparaged owing to its potential for misinterpretation as materialism, pantheism, or other assorted heresies. The passage also announces More's interest in a range of phenomena that, if not altogether undetectable or invisible, are transient, unstable, and elusive. In More's cluttered world, a trumpet is not just a trumpet; it is also a visual field against which a sound occurs. This sound, in turn, requires collateral study, observation, and explanation along with the trumpet, even if the sound becomes undetectable after its brief tenure in the bore. More blazons an interest in what might be called "substantive emptiness," in spaces that are at once empty and yet seem to be filled by something (here, sound). He complicates and enlarges the identity of the trumpet, which is sometimes qualified by its sound and sometimes not, which is sometimes what it appears to be, but sometimes possesses presently undetectable attributes. The trumpet would never be a trumpet if it didn't have sonic attributes now and then, yet the trumpet seems to exist when its sonic attributes escape the ear.

Similar problems with the amplification of identities through fugitive attributes arise with respect to the partitioning of infinity. Cautioning that any segment of infinity is also infinite, math-loving More celebrates heroical fractions in Spenserian stanzas:

What sober man will dare once to avouch
An infinite number of dispersed starres?
This one absurdity will make him crouch
And eat his words: Division nought impairs
The former whole, nor her augments that spares.
Strike every tenth out, that which doth remain,

---

[6] Henry More, *Democritus Platonissans*, in *A Platonick Song of the Soul*, ed. Alexander Jacob, st. 3, p. 406.

An equall number with the former shares,
And let the tenth alone, th' whole nought doth gain,
For infinite to infinite is ever the same.[7]

The paradox of fractional infinity appeals to More because it increases the sum of paraphernalia in the world, because it seems to make something out of nothing and thereby to lend substantiality to space, vacuum, and void. One infinity is nice, but to take a piece of that infinity and find within it another, previously undetected infinity is better still. Stanzas 44 and 45 take a further step into the invisible world by expounding this same paradox with respect to pure numbers rather than tangible planets: "Each part denominate doth still abide / An infinite portion" as infinities spring from infinities in the abstracted world of arithmetic. More's pre-Newtonian algebra allows no countervailing notion of "infinitesimals" whose infinite smallness counterbalances their infinite numbers; rather, More's fractional infinities constitute new infinities. These infinite subsets of infinity even insinuate an ontological claim, a suggestion that all these infinities must exist somewhere.

The far-reaching ontological force of such claims becomes clear when More alleges that even "emptinesse" has extension, that emptiness is not simple, vacuous void but genuine "distance," actual experiential space that travelers could traverse. More adduces an æsthetic criterion within his speculations through which to increase the substantiality of the infinite spaces that his poetry creates. "Distance without end" is open for realtor development; "this unbounded voidnesse" is no mere nullity but rather an implicit affirmation that "The number thus of th' greatest that doth fit / This infinite void space is likewise infinite."[8] Into the void comes a new and promising criterion of *fitness*, both in the sense of adequate seating capacity for a large number of planets and in the more purposive, more æsthetic sense

---

[7] *Democritus Platonissans*, st. 28, p. 413.
[8] *Democritus Platonissans*, st. 42, pp. 417–18.

of artistic propriety. All this vast space would go to waste were it not filled up with fitting worlds, worlds that are invited into being less on evidentiary grounds than because they literally fit into the cosmos and because it is æsthetically fit that they festoon space. Just as More's trumpet provides a fit field upon which sound must emerge as an insubstantial attribute, so deep space provides a field into which worlds must insert themselves. At the very least, seemingly empty space is surprisingly articulate, for it contains and declares information about the locations where new worlds could be situated. Æsthetics, ontology, and science converge as fitness provides an ad hoc explanation both for the presence of planets and for the fit activities (such as orbiting through space) that planets do.

More regards space as *æsthetically* constitutive. Space is far more than a large hole into which God deposits an assortment of gases, liquids, and minerals. Rather, space is an elastic if invisible field of æsthetic potentiality, a blankness that invites and "fits" whatever might erupt in it. The pause that refreshes, space is empty and exempt from sensory detection. Yet, despite its paradoxically "apparent" invisibility, it both provides a pleasing intermission between objects or worlds and attracts, in its exquisite fitness, further filling. More mentions but never affirms the popular seventeenth-century theory that a subtle medium of æther pervades "extension," nor does he subscribe to the Spinozan notion that the universe is an absolutely full plenum, worrying that a universe full in every nook and cranny would render motion, change, and creation impossible. For More, creation requires a receptacle into which to flow, an empty and yet highly productive space to fill. In effect, More switches the emphasis in the idea of the "plenum" from the contents filling the universe to God's unending creative act, from ontology to process. The question is not whether the universe is absolutely full, but whether its open spaces allow room for a temporally unlimited process of creation. In his more esoteric and elusive moments, More indicates that invisible space bends and stretches toward visibility, that it somehow or other—More never makes clear exactly how—shows and declares itself eligible territory for those infinitely numerous yet intermittently placed planets, stars, and other paraphernalia that mark it out and make it to some degree visible. A Platonist but not a prophet, More has no inkling of the probabilistic, chaological, quantum physics that would come to dominate our own post-Heisenbergian day, yet for Hermetic reasons he presents a space that is always on a quantum brink, on the edge of attracting some event or on the point of consolidating into objects, gravity wells, stars, planets, or who-knows-what-next.

The eligibility of More's elastic space for partial filling by "fit" objects is, as presenters of academic conference papers like to say, "part of a larger project": to wit, More's plan to make multiple infinities congruent, to find a way to fit the infinity of worlds within the infinity of space. In a bewildering series of stanzas More describes a universe shaped like two asymptotic cones. The bases of these infinitely extended cones are the Creator while their tips are the created world and the individual viewer. Geometrically complex, these cones appear to have infinite extension and magnitude in all directions when viewed from any point upon them.

> Each portion of the *Cuspis* of the Cone
> (Whose nature is elsewhere more clearly shone)
> I mean each globe, whether of glaring light
> Or else opake, of which the earth is one.
> If circulation could them well transmit
> Numbers infinite of each would strike our 'stonished sight;...
> A circle whose circumference no where
> Is circumscrib'd, whose Centre's each where set,
> But the low Cusp's a figure circular,
> Whose compase is ybound, but centre's every where.

Whether viewers look in the direction of creation or look in the direction of God, they look out on either God's infinity or the infinite remainder of the universe.[9] This double asymptotic cone is so geometrically complex that it not only makes room for but also fits anything, no matter how wryly configured.

So genial a subdivision of space, one allowing accommodation for anything, would seem to answer More's questions about the capacity of the cosmos. Unexpectedly, it opens another set of literary-scientific questions. Working long before the formulation of the inverse square law or the discovery of the Doppler Effect, More, confronted with an infinite number of celestial objects, wonders why the universe is not blazing with light. If there were an infinite number of luminaries in any direction, it would seem that the heavens should sizzle with surplus brilliance. Any one point in the heavens would be backed up by an infinite quantity of stars and starlight—all the more so if invisible, transparent space lacks sunblocking power. To further complicate matters, More wonders why it is that, at night, dark is so neatly proportional to light. Why is there is not one dazzling corner of space where all the stars are clustered, and another where total darkness reigns? Why does the universe evidence neoclassical symmetry and proportion? Unlike many other advocates of the argument from design, More asks not why the world *works* so exquisitely well, but why it *looks* so nice, why it evidences superior compositional qualities. More goes Milton one better by detailing the creation of an invisible yet proportionally opaque "Night":

> This Endlesse large Extent...
> Was at first all dark, till in this spacious Hall
> Hideous through silent horrour torches clear
> And lamping lights bright shining over all
> Were set up in due distances proportionall.
>
> Innumerable numbers of fair Lamps
> Were rightly ranged in this hollow hole,
> To warm the world and chace the shady damps

[9] See *Democritus Platonissans*, sts. 7–8, p. 407.

Of immense darknesse, rend her pitchie stole
Into short rags more dustie dimme then coal.
Which pieces then in severall were cast
(Abhorred relicks of that vesture foul)
Upon the Globes that round those torches trac'd,
Which still fast on them stick for all they run so fast.

Such an one is that which mortall men call Night
A little shred of that unbounded shade.[10]

More characterizes the universe as an amalgam of positive opacities, of rag-like shreds of darkness woven around sundry planets, thereby forming a pleasing pattern likeunto the patterned lattice-work mentioned in Kant's *Critique of Æsthetic Reason*. Earth's diurnal rotation into darkness allows us a recurring pass alongside one such piece of the dark-light lattice-work. Such shards of invisibility play a clearly positive, æsthetic role by raising invisibility, opacity, and other negative, obstructive qualities to prestigious positive status. More frequently relies on rhetorical redundancies such as "hollow hole" to play up the positive, visible aspects of invisible darkness, even despite the fact that his Platonic world-view precludes granting positive existence to privations, here privations of light. Somewhat circularly, More uses his neo-classical ideas about the balanced arrangement of stars and darkness to justify the positive existence of void spaces. He takes the Palladian as well as utilitarian stance that stars are placed at regular intervals, that they are set up to provide the perfect level of minimum night-time illumination. Something, even a negative something like darkness or invisibility, is needed to insure this decorous and decorative spacing, damping, and buffering. More thus gives God reasons to leave some spaces open: to secure this perfect balance of light and dark and to please the inhabitants of various planets with refreshing rotations into partial dark, into the classic, Van Gogh-style "starry starry night."

---

[10] *Democritus Platonissans*, sts. 17–19, p. 410.

One procedure implicit in empiricism from Sir Francis Bacon onward is *extrapolation*, the use of local or limited data to characterize similar things and events elsewhere. If water freezes at 32° Fahrenheit in Chile, most Baconian scientists would assume that it freezes at 32° Fahrenheit in Morocco. This inductive, extrapolative method is phenomenon-specific. When scientists determine the freezing point of water, they edit out extraneous complicating issues such as differing contaminants that might alter its phase-change point in different venues. Modern science postulates a pure, perfect, and universal water that behaves the same way everywhere, much as aviators postulate, for the sake of calculating aircraft performance tables, an idealized standard day of 15° Celsius at 1013 millibars of pressure, a day rarely if ever encountered. Early empirically-minded thinkers had few such limiting or idealizing procedures in place. Underlying Cowley's and Sprat's suggestion that traveling professors should collect data from various locations worldwide is the suspicion that different environments might indeed alter even a basic datum like the aforementioned freezing point.

More tries everything possible to complicate rather than to universalize scientific extrapolation. He *always* takes into account a myriad of modifying factors, influences, and environmental circumstances. His divagation on the darkness that surrounds the "torches" in the "spacious Hall" of the universe extrapolates not only from stellar phenomena but also from their surroundings, arguing not only that "if there is one sun, there must be many," but also that "if our sun is surrounded by a shard of darkness, there must be plenty of other shards of darkness floating around star-spangled space." To make the matter clearer by analogy, were More studying the freezing point of water, he would not only extrapolate the temperature at which water freezes in England to determine the freezing point of water in all other locations, but would also affirm that every ocean has seaweed, that ducks float on every pond, and, given his interest in invisibility and negativity, that every planet must have deep dark holes called "oceans" or "lakes" into which freezable water can be dispensed. If there is any novelty or singularity associated with our planet, we should assume that other planets come bundled with novelties and singularities—or, more skeptically but optimistically, we should assume that there is a limit to the number of truly universal scientific laws, a limit that downplays the similarities of worlds in favor of emphasizing local singularities and that thereby tacitly promises the eventual discovery of infinite variety in the cosmos. Invisibilitarian More thus practices a counterextrapolation in which he enumerates what *blocks* our sight of far-away phenomena as a way of conjecturing what we *would* likely see in other worlds. Just as there are solid objects on the earth that block the view of objects behind them, so there must be innumerable visibility-obscuring objects out in space.[11] If there is an infinity of worlds to see, there is also an infinity of worlds to get in the way. Each of these worlds, whether blocking or visible, will have its shares of novelties and singularities. The less we see from some one point, therefore, the more there must be to see elsewhere. Even if the dispersal of objects through space mediates this sunblocking effect, the abundance of objects in the heavens still means that we need more of a periscope than a telescope to peer around the maze of celestial obstructions than a telescope to look straight at them. More is less naive than descriptive poets like Denham, Dyer, or Pomfret; these bards describe everything they see, but More also tries to describe everything he *can't* or *won't* see and to show how these invisibilities might effect investigative procedures. More warns that he argues not for the infinity of *this* world, of earth and its precincts, but rather for the infinity of world*s*, in the emphatic plural.

Always aware of particular variations, More wants his narrative science to progress from this world to others. This expansion of science is by no means a matter of direct extrapolation, but rather a probing of convoluted detours; even earthbound science reminds us that empirical investigation is always running

---

[11] See *Democritus Platonissans*, st. 12, p. 409.

up against complications and obstacles.[12] Any student of empiricism quickly discovers that Locke, Berkeley, Hume, and most of the lesser-known empiricists spend a lot of time composing nonempirical rhapsodies, rapturously praising universal orders that could never have been directly detected using either unaided sense or Augustan-era instrumentation. More's poetics of obstruction, darkness, and counterextrapolation exposes the reliance of early scientific writers on this literally dark side of enlightened science, on the parts of experience that are obscured by foreground intuitions. More licenses scientific imagination by highlighting the degree of darkness, obscurity, and invisibility in the common experience that empiricism allegedly prizes. He presents a hypothetical attempt to dig to the center of the earth as a metaphor for proper imaginative extrapolation about the unseen fringes of the cosmos. Which "wight," he wonders,

> Did dare for to invade
> Her [Earth's] bowels but one mile in dampish shade?
> Yet I'll be bold to say that few or none
> But deem this globe even to the bottome made
> Of solid earth, and that her nature's one
> Throughout, though plane experience hath it never shown.[13]

Sweaty and beleaguered, we soon hit a point where we will surrender the investigation of the earth to moles and earthmovers, concluding that obscurity and obduracy can become extrapolable counter-information.

Modesty-free in his boast that, through the "inward triumph" of scientific imagination, his "soul" is "stepping on from starre to starre / Swifter than lightning" while "Measuring th' unbounded Heavens and wastfull skie,"[14] More hardly qualifies as the Green Party poet-of-choice. However much he likes wild or open areas, he still sees unused space as a waste. Whether he means "wast-

---

[12] See *Democritus Platonissans*, st. 21, p. 411
[13] *Democritus Platonissans*, st. 62, p. 423.
[14] *Democritus Platonissans*, st. 5, p. 406.

full" in the obsolete sense of a "desert" or in the modern sense of "underutilized," More treats space in a doubly-negative way, as a negation that should be negated into a poetic something. More's centrifugal verse tends to run off into vastnesses and fly off into tangents, to convert encounters with nothing into opportunities to visit substantial (if invisible) worlds. Dispersive and centrifugal, More's rhetoric is designed to stretch beyond any limits and to elasticize any boundaries, invisible or visible, that he may confront. Any one big thing, whether ontological or poetical, will be broken up into an infinity of smaller units, whether neat stanzas or solar systems. The passage cited above, on the dispersion of darkness through and around the universe, is but one example of More's intolerance for any one single large mass or volume. If More encounters any such phenomenon, he breaks it up into bits and pieces, then scatters those bits and pieces into the expanding vastnesses of space. Infinity for More is always a vast collection of positive and negative, light and dark, opaque and transparent, and visible and invisible shards and shreds. In More's Spandex view of cosmology, light itself needs to be parceled and stretched out into far-flung bits and pieces so that it won't overwhelm any one volume of space. "Nature and carefull Providence...did so contrive" that

> Th' Hearts or Centres in the wide world pight
> Should such a distance each to other bear,
> That the dull Planets with collated light
> By neighbor suns might cheared be in dampish night.[15]

"Chear," a psychological and æsthetic commodity, arises from the dispersal of large, concentratedly hot objects to due distances, from which remote stations they maximize the enjoyment of life on the myriad planets throughout the depths of space. Egalitarian More reminds us, in the next stanza, that our own sun has been similarly dispersed, becoming "a starre elsewhere" where it

---

[15] *Democritus Platonissans*, st. 23, p. 412.

"cheareth" "those dim duskish Orbs" that must "round other suns" "run."[16] Beauty, cheer, and the æsthetic appeal of the universe rest on the extenuated foundation of dispersal, infinity being æsthetically at its best when stretched out over vast infinities. More's sprawling verse recapitulates this cosmological process by dispersing his ideas over vast epic-poetic space such as those carved out by the Spenserian stanza, More's poetic measure of choice. More's poetic recapitulation of vastness may be difficult to appreciate in our impatient, brevity-minded time, but in More's neo-Ciceronian rhetorical context it would surely have been lauded as a fulfilment of both a dispersive rhetoric and a dispersive ontology.

The almost childlike pleasure that More takes in throwing confetti-like planets into space interacts with his ontological understanding of æsthetic "fit" and "fitness." In a stunning feat of elliptical reasoning, More, who affirms the eternity as well as the infinity of worlds, shortcircuits the need for a genesis myth. Empty space has an æsthetical and generative elasticity, invoking various types of creations out of its own receptivity. A truly empty space is inconsistent with its invitational character:

> Else infinite darknesse were in this great Hall
> Of th' endlesse Universe; For nothing finite
> Could put that immense shadow unto flight.
> But if that infinite Suns we shall admit,
> Then infinite worlds follow in reason right.
> For every Sun with Planets must be fit,
> And have some mark for his farre-shining shafts to hit.[17]

Reconstructing More's reasoning is always a challenge, but here he seems to regard the notion that a space that could accommodate solar systems might not do so to be flatly intolerable. Possibility slips into necessity as More assumes that a space that *could* house

---

[16] *Democritus Platonissans*, st. 24, p. 412.
[17] *Democritus Platonissans*, st. 26, p. 413.

planets *should* or even *must* do so. By teleological extension, suns with radiant warmth require duller planets as targets for their infrared benevolence. We learn that planets, owing to their gravitational generosity, require moons.[18] Fitness and needfulness seem to have ontological implications in More's highly sociable universe, where planets who need planets are the luckiest planets in the world. It isn't long before More unveils a lavish descriptive passage about all the social and ecological systems that are required by the inviting surfaces of all these worlds,[19] and not long after that before More concludes that vacant, elastic, and inviting space neither could nor would have blocked the Creator's mighty efflux.[20] More is attempting to have his ancient ontology both ways, to present space at once as a primeval void that elicited Creation and also as a Hermetic-Zoroastrian negative entity that potentially could oppose the Creator.

More's poetry, then, is energized and to some extent organized by a huge field of space, darkness, and invisibility. The full exposition of More's eschatology of the infinite cosmos is a topic for another paper, but in general More's account of the goal and end of the world combines egalitarianism with diversification in a celebration of the elasticities and potentialities of nullity. More holds that no segment of empty space, even that little corner of the universe that once comprised the void into which our earth would later condense, enjoys any privilege over another. There is no reason to believe that the space that now contains earth is unusually fruitful. The fact that our earth erupted in the spot that we now occupy proves that *any* bit of space can both generate and accommodate vastly complex things. Purposive, space seeks to house "worlds" and all their dramatic stories. We may assume that any and all parts of space are generating world after world, all of them abounding in dazzling variety. At the same time, renewable space, vacuity, and nullity are necessary to accomplish this expression of infinite miscellaneousness. New places for new

---

[18] *Democritus Platonissans*, st. 30, p. 414.
[19] *Democritus Platonissans*, st. 31, p. 414.
[20] *Democritus Platonissans*, st. 48, p. 419.

planets and their evolving stories must keep opening up, even at the expense of old ones. Worlds must have an end characterized by the same crepusculant variety that More finds everywhere else. Meteors, comets, supernovas, and assorted agents of cosmological calamity are key to More's cosmology, so much so that he embraces the patristic philosopher Origen's heresy that dozens of earths, dozens of Adams and Eves, and even Christian soteriological stories, have all come and gone in the colossal recycling bin of our time. Variety is a harsh master; if there is an infinite number and variety of worlds, there must also be an infinite number and variety of world-ending and space-opening calamities. In the last analysis, More's elastic ontology of emptiness stretches beyond its own snapping point. Breaking down and regenerating, coming to the very limits of verse expression and then charging past them into rhetorical explosions—all these cycles are integral parts of More's ever-elastic, ever-elongating, and ever-balanced world[s].

II: John Hutton's Creative Caves

Over a century after Henry More and after numerous visits to the deeper recesses of the Noel Collection, John Hutton surfaces as a leading practitioner of the prose if not poetry of invisibility. Between More and Hutton lie numerous temporal, historical, political, philosophical, and just plain open spaces, yet Hutton seems to prove More's point that spaces, vacancies, and invisibilities generate variety out of similarity and excess out of symmetry. Hutton should count as the climax and the sine qua non of a special variety of Enlightenment writer—whether Henry More, the Graveyard Poets, the William Collinses and the Thomas Grays, or the late-century prophet-poets like Christopher Smart and William Blake—who wrote largely or exclusively about things that no one has yet seen or that can't ever be seen or that otherwise highlight invisibility. Over the last century of criticism, we have come to think of the "pre-Romantic" as well as the Romantic periods as times of vast social and scientific revolution. The term

"Enlightenment" is often appropriated specifically to the glamorous, highly public French Revolution and its rationalist, secular, and political ideologies. Cave, hole, crevasse, and pit explorer John Hutton reminds us that later Augustan culture still had plenty of room for small domestic or cramped dark poetic spaces that evidenced just as much cosmological potential as Henry More's infinite distant worlds.

The now-forgotten Hutton rose to prominence on the strength of his *A Tour to the Caves, in the Environs of Ingleborough and Settle, in the West-Riding of Yorkshire.* Undertaking a tour of underground formations in northeast England, Hutton assumed the formidable task of describing hidden or even inaccessible locations, many of which loomed in complete darkness. Hutton stands apart from most travel writers of the period, whether the high-profile Samuel Johnson, the wide-wandering Mark Catesby, or the numerous anonymous chroniclers of exotic places. These travel writers tend to look at vast or diverting phenomena that are brightened by daylight or illumined by the presence of human social environments, but Hutton probes what eludes the human eye. True, many travel writers, including Johnson in Scotland, now and then visited a cavern or hole or pit, but often only to demythologize it, as when Dr. Johnson used his measuring stick to scale down the romantic legends associated with a Scottish cavern. Hutton, on the contrary, does all that he can to play up the elusiveness, inaccessibility, and invisibility of his underground environments. Hutton is the most lavish and "purest" of a large cadre of cave-connected writers, whose reflections on underground environments I will be exploring in another essay. But while these many writers, spelunkers, and explorers tend to focus on the inhabitants or the geology of the dangers of caves, Hutton attends primarily to their darkness and invisibility, to the suggestive if negative æsthetic power of their insensibility, to the empirical study of our reactions to the *lack* of sense data, and to the ability of the invisible worlds beneath the earth's surface to inspire a body of prose compositions.

It would be easy enough to say that Hutton is simply imbibing the spirit of early Romanticism or is on a typical late-century quest after the sublime. Unfortunately, there is too much of the baroque left in Hutton for so Puritan or psychological a project. Rather, Hutton applies himself to the literary construction of invisibility, linking it to familiar, illuminating figures from literary classics in much the same way that More sought to associate dark space with Spenserian archaism.

> Having never been in a cave before, a thousand ideas, which had been for many years dormant, were excited in my imagination on my entrance into this gloomy cavern. Several passages out of *Ovid's Metamorphoses*, *Virgil*, and other classics crowded into my mind together. At one time I thought it like the den where *Cadmus* met the huge serpent...Indeed there wanted nothing but an ancient wood, to make one believe that *Ovid* had taken from hence his lively description. As we advanced within it, the den of *Cacus* and the cave of *Poliphemus* came into my mind....The light we had seemed only darkness visible.[21]

The deeper and darker Hutton goes, the faster and brighter come the allusions, culminating in a Miltonism that caps the miniature history of literature transcribed in this dark cave. By diverting from the cave into an imaginative literary projection of it, tricky Hutton sets up an association, like Henry More's, between darkness and proliferation, between invisibility and abundance—here, an abundance of familiar, culturally visible literary allusions.

---

[21] John Hutton, *A Tour to the Caves, in the Environs of Ingleborough and Settle, in the West-Riding of Yorkshire. With some philosophical conjectures of the deluge, remarks on the origin of the fountains, and observations on the ascent and descent of vapours, occasioned by facts peculiar to the places visited. Also a large glossary of old and original Words made use of in common conversation in the North of England. In a letter to a friend.* (London: 2nd edition, 1781), 14–15.

It is also tempting to attribute to Hutton the same sort of artful artlessness that we find in Romantic poets, poets who, despite their alleged allegiance to nature, take every opportunity to allude to John Milton, the Ancients, or anyone who dwells in a library rather than a forest or field. The difficulty with this interpretation is that Hutton is not citing classical or even contemporary sources for his descriptions, but rather is rejecting visible nature in favor of describing almost nothing. Giving intelligible form to vacuities that elude observation, Hutton follows the procedure of Henry More rather than that of Blake, Keats, or Wordsworth; he keys his encounters with darkness not to this or that "minute particular" or to this or that sense impression, but rather to tales and anecdotes, to local legends and stories that extend, enlarge, enhance, multiply, and otherwise infinitize the worlds that surround whatever he is talking about. Put oxymoronically, anything that Hutton cannot see ends up inviting, eliciting, or getting attached to some or other story, anecdote, or observation. His procedure recalls that of Henry Fielding, who, whenever he encounters something unknown, spends less time reporting the encounter than in attaching an interpolated tale to it. Although Hutton sees nothing of the dark cave, he receives certain credible reports about incredible doings in it.

> He [the guide] mentioned two circumstances we paid some attention to. About fifty or sixty years ago, a madman escaped from his friends at or near *Ingleton*, and lived here a week, in the winter season, having had the precaution to take off a cheese and some other provisions to his subterranean hermitage. As there was snow on the ground, he had the cunning of *Cacus* (see *Virgil, Æneid*, L. 8, l. 209) to pull of the heels of his shoes, and set them on inverted at the toes, to prevent being traced: An instance,

Figure 1: Entrance to Ingleborough Cave, site of John Hutton's explorations, now open to the public as a tourist attraction.

among many others, of a madman's reasoning justly on
some detached part of an absurd plan or hypothesis.[22]

Additional gruesome tales follow, with the void space of the
cavern being used as a setting for miniature histories, as the focus
around which an infinity of tiny narrative worlds turns. Hutton
is a great one for the extensive description of detailed, even
microscopic abundance—he talks of "crane's bills, scurvy-grass,
birds' eyes, various liver-worts, orchises, row-wort, lilly of the
valley"[23] along with dozens upon dozens of berries, fruits, lichens,
mosses, and plants—but he does so less to celebrate concrete detail
than, like More, to substantiate vacuity by surrounding it with
collateral detail. Hutton resembles the great visual explorers of
America—illustrators and narrators like Mark Catesby, Zadok
Cramer, and John James Audubon—in his desire to compensate for
the vast unknownness of America by filling in smaller spaces, by
penciling in every feather on every bird or recording every pismire
under every stone. In point of fact, the greatest portion of any
nation, whether America, Britain, or elsewhere, cannot, for
practical reasons, be seen. No one can tour every possible site at
every level, from cave to cloud or from Cornwall to Orkney.
Maps, gazettes, and travel accounts record not only what a few
have seen, but also what most will never see and what no one can
hope to see in totality. By reporting on the unknown and
inaccessible, exploration literature like Hutton's conflates the
invisible with the observed and recruits the unperceived into the
telling of tales and the creation of variety. Not being present and
visible to be judged, anything that Hutton describes is free to
become appealing, interesting, sublime, or romantic. For all a
reader knows, a "hurtle-berry or bill berry" or any of the other new
wild delicacies mentioned in Hutton's account might have a horrid
savor; yet, so long as this distant berry is wrapped up in More-like
shreds of invisibility, it remains free of judgment and constitutive

---

[22] *A Tour to the Caves*, 17–18.
[23] *A Tour to the Caves*, 32–33.

of poetically evocative experience.   Like Henry More, Hutton rushes into an æstheticized and cosmologized version of latitudinarian-libertarian ideology. Everything should be tolerated or even applauded simply because it is various—a mentality that science today has tamed into popular environmentalism, with its catch-phrases about the immense ecological variety of the Amazon basin or the rich resources of arctic wildlife reserves. Hutton can even tolerate hard-core monastic Catholicism, at least so long as some of its detail is obscured by time and thereby rendered invisible. Horrified at the post-Reformation ruination of Kirkstal Abbey, Hutton admits grudging respect for the monks who built it.

> We stood some minutes looking with silent respect and reverence on the havock [sic] which had been made by time on this sacred edifice. How much soever we might condemn the mistaken notions of monkish piety, that induced the devotees to forsake all the social duties of life in order to be good men; yet we secretly revered that holy zeal which inspirited them to exert every power in erecting structures, the magnitude and beauty of which might excite ideas worthy of the Deity to whom they were dedicated.[24]

Magnitude and beauty, novelty on a grand scale, warrants attention if not applause. What saves the abbey from the darkness of superstition is what saves the cave from invisibility: the ability to localize a vast array of attendant anecdotes and tales, in this case Hutton's musings on abuses by religious reformers. A ruined abbey, like a dark cave—and like an unknown world in the dark chasms of space—is a focus for the study of an infinite variety that we, in our finitude, are never going to see fully, a variety that abides somewhere between literal invisibility and conceptual demonstrability.

[24] *A Tour to the Caves*, 55.

What is perhaps most remarkable about Hutton is that he gives his anecdotal reports a technical spin. He is not only interested in telling sensational tales about escaped madmen, dangerous *banditti*, and unhappy victims, but he is also enamored of the specifically scientific imagination. It is important for Hutton to understand precisely how much cheese and what kind of shoes were needed by an escaped madman. The combined imaginative-technical appeal of the Ingleborough caves is that they support a huge range of speculative scientific theories about anecdotal and highly various phenomena. Hutton spends pages both hypothesizing about and reporting local folklore concerning the geyser-like well at Kirkby-Lonsdale.

> We turned to the right, along the road toward *Kirkby-Lonsdale*, about a mile under the high and romantic rocks called *Giggleswickscar*; in order to see the well by the way side, that ebbs and flows. We were in luck seeing it reciprocate several times while we were there, and not staying above an hour. We could not however learn, with any degree of certainty, by what intervals of time, and to what heights and depths the reciprocation was carried on. We were informed that if the weather was either very droughty or very wet, the phænomenon ceased. I have seen some philosophical attempts to solve this extraordinary curiosity on the principle of the syphon, but in vain, as on that hypothesis, if the syphon is filled by the spring, it will flow on uniformly for ever. We are told by drunked *Barnaby*, a hundred and fifty years ago, that it puzzled the wits of his age...

> > Thence to *Gigglewich* most steril,
> > Hem'd with shelves and rocks of peril,
> > Near to th' way, as a traveller goes,
> > A fine fresh spring both ebbs and flows;
> > Neither know the learn'd that travel,
> > What procures it, salt or gravel.

> Two country gentlemen, about thirty or forty years ago, promised something more successful in the issue of a paper war that was carried on between them, to the great amusement of the neighbourhood: Nothing however was determined or contended for about this well, so famous in history, but whether it was a natural curiosity or not.[25]

The stories resulting from this geological study are a remarkable mixture of pop science, fairy tale, cutting-edge research, and local color. They are, in sum, the literary equivalent of ecosystems, of worlds within worlds. In yet another, still more painterly passage, Hutton parlays the long delay between the dripping of water from overhead rocks and the audible plunking of droplets into an underground lake with an anecdotal report that black trout might swim in the black abyss of that subterranean pool, all to create the illusion of tremendous depth.[26] Science manufactures as well as reports on the sublime as independent details focus into a verbal projection of deep underground space.

It should come as no surprise that the quest for variety in the midst of spacious darkness should lead Hutton to the same sort of apocalypticism that we found in Henry More. Through twisted æsthetical-scientific reasonings that range from dark English caves to the blinding white convexities of Gibraltar, Hutton defends his odd notion that Adam's fall and Noah's deluge were followed by cataclysmic upheavals that built mountains through the scattering of bones.

> The rock at *Gibralter*, and several mountains in *Dalmatia*, and no doubt, many others in different parts of the world, are made up of bones, not only of every animal extant in nature, but particularly those of the human species.[27]

[25] See *A Tour of the Caves*, 47–48.
[26] See *A Tour to the Caves*, 23–24.
[27] *A Tour to the Caves*, 55–60.

There follows a lurid account of the disasters that must have occurred, either when a comet struck our planet or when the primeval earth was wound up into its present 24-hour rotation. At the least, Hutton's theory of the evacuation of bones from the earth to make mountains is original and dramatic. It is surely one of the very few examples of the literary use of cataclysmic seismic inversion, in which buried or underground or otherwise concelaed remains are ripped up wholesale and piled into convex, elevated, exposed, and visible natural monuments (only Dante's account of the erection of Purgatory as a result of Lucifer's excavating plunge into the infernal pit could rival it). Such sublime drama was one requirement for science during the Enlightenment, when the origins and the ends of the world were favorite topics among the neo-Puritan gentlemen who met at Wadham College and their followers and successors. This notion that science must have astounding implications persists today in our belief that earthshaking discoveries are more rewarding than basic research, that it is perhaps better to unriddle the secrets of a volcano or to plumb a black hole in space than to figure out how to improve drainage around Shreveport.

III: George Friedrich Meier, Comedian-Professor

One afterthought to emerge from a study of Henry More and John Hutton is that every scientific or scholarly paper ought to have a coda. Given the combined cosmological and subterranean momentum toward variety and multiplicity that these gentlemen-researchers report, no paper should end without one more anecdote emerging from it. The phrase used on every grant application, "more research is needed," loaded as it is with suggestions of incompleteness and incompetence, may have its origins in the relentless quest for multiplicity that spans early-modern science. Similar observations can be made about the tone of scientific and scholarly exchange. Both More and Hutton occasionally diverge into humor, satire, and jesting, a ribald inclination that, in our serious-minded, post-romantic research milieu, nowadays more often appears in the enthusiasms of popular science than in academic jesting. The lessfiltered eighteenth century was eager to make it both a part of and a topic for science—which brings us to Georg Friedrich Meier.

A mid-century Prussian philosopher and scientist, Meier dealt with an assortment of offbeat topics, topics that seldom made their way into academic study because they were either lacking in dignity or resistant to the visually-attuned empirical science of the day. Dubbed the "Merry Philosopher," Meier set himself up not as a Bergsonian prophet of comic irony but as a practitioner of the science of jesting. For this, he took more than a little collegial heat:

> Several Exceptions were made by some formal gloomy persons to these Thoughts, on their first Publication: they account the Undertaking, indecent and ridiculous: they imagined I set up for a Professor of Jesting, and publicly declared, I affected an extraordinary Turn that way, and wanted to keep it in Exercise....I only wanted to be thought a Whet-stone for sharpening Iron, without pretending to cut.[28]

Issuing harsh warnings against university professors who try to make their lectures appealing to students through the heavy use of bad jokes—"Publick Professors in Universities often disgrace themselves by wretched Jests, with a View to divert their Hearers, and relieve the Severity of the profound Truths they are proposing, by interlarded Jests"[29]—Meier announces his plan to present a

---

[28] George Frederick Meier, *The Merry Philosopher; Or, Thoughts on Jesting. Containing Rules by which a proper judgments of JESTS may be formed; and the Criterion for distinguishing TRUE and GENUINE WIT from that which is False and Spurious: Together with Instructions for improving the Taste of those, who have a natural Turn for Peasantry and good Humor.* (London: 1765), 2–3. Meier's work was rendered into English by an anonymous translator.

[29] *Merry Philosopher*, 7.

scientific account of how humor should be formed and likewise to install in his audience a technology for judging wit.

Meier's whetstone image admits a degree of imprecision. Meier recognizes that one cannot cut through the irregular topic of humor in the same way that one might parse the parts of a planetary orbit. Many eighteenth-century writers—Alexander Pope, Abel Boyer, Joseph Addison, John Dryden, and Immanuel Kant, to name but a few—had theories of humor that treated "the comic" as a positive phenomenon and that applied traditional analytic and discursive tools to solve its mysteries. Meier takes a pioneering approach to the topic by following the example of the English virtuosi and proposing a highly empirical, locally anecdotal study, by looking first at particular jests and then extrapolating to broader issues related to humor in general. Humor, in turn, is portrayed as a kind of elusive master science, as a *je ne sais quoi* that evidences few positive attributes but that opens many opportunities for error. Meier congratulates himself in having gone one step beyond conventional science to develop a new protocol suited for the study of imprecision.

> The Æsthetic is a Science, which in general treats of our sensitive Knowledge. It examines into the Perfections and Imperfections of this sensitive Knowledge, and of this Expression: It deduces the Rules, by the observation of which the former may be obtained, and the latter avoided; It shews in what Manner the inferior cognoscitive Faculties are to be improved and employed, in order both to speak finely: And it supplies us with the Grounds, or Principles, by which we may in a just and rational Manner pass a Judgment on the several species of sensitive Knowledge and the Expression thereof.[30]

Meier's theory has all the scaffolding and positivistic confidence of a scientific undertaking; he promises to "enable Persons with a

---

[30] *Merry Philosopher*, 12.

Turn for Jests, to distinguish the false and insipid from the genuine and sprightly; to stifle in the Birth all low and indecent Drollery; to repress imprudent Sallies of Wit, which spare not even a Bosom-friend; to prune the luxuriancies of a wild Imagination, Faults, the Wittiest and most Ingenious may at Times be subject to."[31]   Meier even unveils a tabular accounting of the "seven cardinal rules" for making good jests: "1. A jest must contain, or excite a proper variety of ideas. 2. A jest must be sufficiently grand, or be an important, fruitful, decent thought. 3. Must be a just thought. 4. A lively, spritely thought. 5. Have a proper degree of certainty. 6. Must be sufficiently striking. 7. Be expressed in a fine manner."[32] Yet whenever Meier comes to the analysis of a joke or even of humor in general, the confident tone of the positivistic scientist gives way to a more cautious (albeit more entertaining) vacillation between the claims of solid rhetorical rules and the whimsies, tangents, and hidden unexpectancies into which any particular witticism takes us.   Thus, Meier initiates a careful distinction between "jests" and "jocose speeches," but before long he has slipped into a joking digression that at once proves, superannuates, and reaches far beyond the axiom that he set out to prove:

> I distinguish jests from jesting or jocose speeches; a jest is the thought unexpressed; and to express the jest happily, peculiar rules are requisite. Diogenes the Cynick coming once to a very small, inconsiderable town, with very large and magnificent gates, he told the inhabitants "to shut their gates, lest the town should run out." This jest or banter implies a jest, ingeniously laying open the ridicule arising from a gate too large for a town so small. Lewis XIV. observing the courtiers riding full speed one after the other; the foremost with an uncommon long chin, the hindmost with scarce any at all; the King asked

---

[31] *Merry Philosopher*, 7.
[32] *Merry Philosopher*, 48.

whither they were driving at such speed? M. de Cleram-
bault replied: "The hindmost is in pursuit of the fore-
most, to recover his stolen chin." Here it is again evident,
that the ridiculous sentiment, *viz.* a person with little or
no chin, pursuing another with a very long one, is
ingeniously represented by a comparison with a theft.
And thus I hope, I have sufficiently cleared my definition
of a jest.[33]

Meier keeps coming back to the posture of a scientist and philosopher, whether by attempting a rather strained distinction between
"banter" and a "jest" or by asserting at the end of his comic
digression that he has indeed vindicated his proposed distinction
between "jesting" and "jocose speeches." The proof presented,
however, is at best indirect, a detour on the path to a science of
comedy. Meier develops a science of surprise in which piquant
tales regularly manifest the sublimity and refinement of humor, by
which the science of jesting resolves into a grace beyond the reach
of art that lays down rules but that itself remains partially
shadowed—partially invisible.

Although his subject abides in the bright space of the coffee
house, the cocktail party, the gala reception, and the sociable
conversation, Meier deals in darkness and vacuity. His jovial
science centers on an imprecise art and admits that humor is the
domain of the lower, less disciplined and less accomplished, less
understandable faculties. Meier plans to give a scientific account
of what we *cannot* know about these elusive and underachieving
faculties; he plans to categorize all the varieties of high and low
humor so as to show how the aforementioned low faculties can
produce an unending variety of jokes. Meier cites Cicero to the
effect that humor is a whole *Gestalt*, that it works in a context in
which a joke hits or misses its mark owing in part to environmental influences.[34] Meier thus creates a kind of "ongoing research

---

[33] *Merry Philosopher,* 37–38.
[34] See *Merry Philosopher,* 22–23.

machine" in which he continually explains why certain jokes work in this or that situation but not in others—a machine that is housed in a huge epistemic cavern, that abides in an imprecise risible faculty whose nature is both above and below our ken. "A jest may please one, which displeases another; make one laugh, while another keeps his countenance....when I commend any thing in a jest, I am not to be understood to approve the whole of it. A jest has various perfections, not always meeting together in the same jest. A jest, therefore, may in one respect be fine, in another, mean; have more imperfection than perfection. And thus different readers may come to consider the same jest from different points of view."[35]

However imprecise Meier's humorous science, it follows the a trajectory comparable to those offered by the more sober Henry More and John Hutton. On the one hand, Meier has a high degree of concern for empirical detail and verification; on the other hand, he is equally eager to fly off into extra-empirical spaces. Humor is a far different topic than cosmology or speleology, yet the procedures and the emotional resonances of all these lesser Enlightenment sciences—the use of a few details to launch into rhapsodic discussions of invisible spaces, concepts, and phenomena—is much the same. For a man accused of being Germany's jesting professor, Meier is unusually and repeatedly at pains to push the science-jesting analogy and to play up the imprecisions and elaborations found in allegedly empirical research. First, it is important for Meier that a joke be built on the foundation of empirical knowledge. The thoughts underlying a joke must be empirically verifiable, technologically correct, and susceptible of extrapolation into other contexts and worlds. Meier provides an example of a joke with strong empirical foundations:

A person of condition married a very mean woman: another said, "This gentleman has brought a spot of oil on his family." This jest affords much matter for reflec-

[35] *Merry Philosopher*, 27–28.

tion. The whole family is compared to a garment: a spot
of oil coming on it, spreads farther and farther; but the
more it spreads, the less it stains the part of the cloth,
into which it soaks. Just so in unequal marriages: their
immediate children suffer most in their honour, but in
the great grand-children, the stain in the noble blood
gradually wears off.[36]

This joke passes Meier's test because it takes proper account of the
physical action of oil on cloth, because that action is congruent
with the thought underlying the joke, and because the metaphor
can be extrapolated into succeeding generations and contexts
without any diminution of its force or intelligence. The more such
a joke can be extrapolated, the better; the more worlds it spoofs,
the greater its perfection. On the other hand, Meier's lust for
extrapolable jokes takes him, just like More and Hutton, into
cataclysmic, dangerous, or morally intolerable spaces. Because
jesting is a pliant and adaptable phenomenon, Meier tries to
imagine every possible sort of joke, a project that takes him into a
comical netherworld.

Criminal jests are denominated from the duties we
transgress in jesting; and may be atheistical or irreligious,
when against our duty to God....[for example,] A Roman
Catholick Lady praying in solemn manner to a saint for
conversion of her husband, in four days after the husband
died; upon which she said, "How great is the goodness of
this Saint! Who grants me more than I prayed for."
Waving other consideration, this jest is criminal, as it is
contrary to the love a wife owes her husband.[37]

It is the unique property of humor to find out such little worlds,
to probe the social spaces that Enlightenment social theory

---

[36] *Merry Philosopher*, 55–56.
[37] *Merry Philosopher*, 42–43.

overlooked. What seems most surprising even to Meier is that his scientific study of jesting leaves him laughing most heartily at precisely those violations and variations of and upon the norm that scientific criticism and æsthetics would condemn as indecorous or immoral. Humor can thus take us over the border of empiricism and into such marginal, hybrid invisible spaces as those occupied by both moral reprehension and death itself. Meier has more than a few examples of perverse gallows humor:

> We have many examples, of persons jesting at their death, There was a law subsisting formerly in France, that a delinquent under certain circumstances should be pardoned, if he married a common prostitute. A native of Picardy who was to be executed for some capital crime, having ascended the ladder, a prostitute, who was lame, was presented to him; and it was his option to marry her, or to be hanged. After surveying her for a moment, he called out to the Executioner, "Tuck up, tuck up! She limps." This jest, indeed, is uncommonly sprightly, as exhibiting a deformed creature to be a greater evil than hanging. But yet the last moments of our life are a period too important and solemn to admit of jesting and mirth.[38]

In such a joke as this macabre one, humorous science flutters back and forth between the seen and the unseen, between the expressible, the inexpressible, and the tasteless. From the caverns of human sinfulness and the abysms of mortality this joke extracts a flash of wit, makes a sally into the unseen world beyond death, and then draws back into rational analysis, into the kind of chatty, daylight-appropriate scientific banter usually found in Sprat, Burnet, Addison, or even David Hume in his easy-going moments. Humor ends up not only as the object of scientific study but as a tool of scientific inquiry, a device for prying open, rhetorically

[38] *Merry Philosopher*, 85–86.

reaching, and psychologically stretching into previously undisclosed places and disciplines. Meier's theory helps to explain the tonal harmony, in the Enlightenment mind, between satire and science, between the cosmology of a poem like Pope's *Essay on Man* and the heavy admixture of ridicule, comedy, and satire that such a poem contains. Despite being the laughing stock of Enlightenment academe, Meier is an open door on the expansivities as well as the invisibilities of early modern science—a door that can open either to the deep recesses of caves and the inky chasms of epistemological space or onto the witty conversations of the Augustan coffee house.

PART II
THE *DRAMATIS PERSONÆ* OF SCIENCE

SCIENCE, MASCULINITY, AND EMPIRE IN ELIZABETH HAMILTON'S *HINDOO RAJAH*

Peter Walmsley
McMaster University

T he Hindu nobleman Maandaara is distressed about his oldest and best friend, Zaarmilla, Rajah of Almora. Zaarmilla, in the chaotic aftermath of the Rohilla campaign, has discovered on his estate a wounded Briton, Captain Percy of the East India Company. Captain Percy, as he languishes near death, proves himself little short of an angel, wise beyond his years and deeply pious. Zaarmilla and Percy become like brothers, and so awed is the Rajah by the spectacle of English masculinity in all its glory that he burns to leave his native land and visit Britain. Maandaara has heard this and is horrified, convinced that Zaarmilla has been bewitched. He takes it upon himself to convince Zaarmilla by letter that all the English, Percy included, are brutal savages, "impious eaters of blood," and at best petty enchanters in league with "evil genii."[1] For proof, Maandaara offers an account of how, when he was visiting the home of an English Saib in Agra, he witnessed the dark English magic first hand:

> Among several other tricks, [the Saib] made the whole company, consisting of more than twenty persons, lay hold of each other's hands, and form a circle, and then by turning the handle of a little instrument, composed only of metal and glass, but which, I suppose, must have contained the evil spirits obedient to his command; he, all at once, caused such sensations to pass through the arms of the company, as if a sudden stroke had broken the bone, which was not, however, on examination, found to be the least injured....At another time he shut out the piercing light of day, which has always been unfavourable to such practices, and made us behold armies of men, and elephants, and horses, pass before us on the wall....Ships rolled upon the bosom of the deep; and men who appeared wild with distress, and panting in the agony of terror, were exerting themselves to save their lives. This sight of horror drew tears from our eyes; and we burst into exclamations of sorrow. When lo! In a moment, the sun being admitted into the apartment, the scene vanished, and we saw nothing but the hangings which formerly adorned the wall. (102)

This is the first of several scenes of science in Elizabeth Hamilton's highly successful *Translations of the Letters of a Hindoo Rajah,*

[1] Elizabeth Hamilton, *Translations of the Letters of a Hindoo Rajah*, edited by Pamela Perkins and Shannon Russell (Peterborough, Ontario: Broadview Press, 1999), 101. All subsequent citations of this work will indicate the pagination of this edition, incorporated in the text of the essay itself. I thank my fellow Noel conferees, all of whom contributed, either in their direct suggestions or their own work, to the rewriting of this essay, but especially Barbara Benedict, Anna Battigelli, and James Buickerood. I am also very grateful to Paul Wood, Professor of History at the University of Victoria, for helping me track down Henry Moyes, who taught Elizabeth Hamilton natural philosophy. And finally, thanks to Sarah Brophy, Julie McGonegal, and Grace Pollock for their many, invaluable suggestions.

published in 1796.[2] Needless to say, the letter I have been quoting is fictional; *Hindoo Rajah* is a satiric novel in the tradition of Montesquieu's *Persian Letters* (1721) or Goldsmith's *Citizen of the World* (1762), imagining how Britain would look through foreign eyes, and delighting in the comedy of a foreign mind struggling to give some coherent account of the myriad follies and vices of England. That said, *Hindoo Rajah* is also a serious attempt at the ethnography of both India and Britain through the liminal figure of Zaarmilla, who as a rajah is a member, albeit a imaginary one, of what Gayatri Chakravorty Spivak calls "the floating buffer zone of the regional elite-subaltern," the class that provided the British with primary access to Hindu culture and law in the early colonial period.[3] As Claire Grogan has recently argued, Hamilton's novel is daring in its radical blending of the genre of the fantastic oriental tale with that of the factual oriental study.[4] Like Montesquieu and Goldsmith before her, Hamilton recognizes that ideas of self and other are inextricable, that it is in fact only through the eyes of the visiting foreigner that we can begin to see ourselves. And like them again, she seems blithely confident about the task of inventing and giving voice to the other, betraying little of the paralysing anxiety of empire and obsession with race that marks Victorian fiction about the subcontinent. The India question had only just become a question in Hamilton's lifetime, and British curiosity about India had not yet been thoroughly crushed under the weight of imperial self-justification to come. Such curiosity is certainly the driving energy of *Hindoo Rajah*. Hamilton had never been to India and was no master of eastern languages, but she none the less had read a great deal of orientalist scholarship and was deeply invested, emotionally and intellectually, in Britain's role in India. Her brother, Charles Hamilton, was a lieutenant in the East India Company who had served thirteen years in Bengal

[2] There were five editions of *Hindoo Rajah* between 1796 and 1811.
[3] Spivak, "Can the Subaltern Speak?," in C. Nelson and L. Grossberg, eds., *Marxism and the Interpretation of Culture* (Chicago: University of Illinois Press, 1988), 285.
[4] Claire Grogan, "Crossing Genre, Gender and Race in Elizabeth Hamilton's *Translations of the Letters of a Hindoo Rajah* (1796)," *Studies in the Novel* 34 (2002): 21–42.

under Warren Hastings and had participated in Hasting's dubious war against the Afghan Rohillas, a campaign of which he wrote an exculpatory history published in 1787. Charles had returned to London in 1786, commissioned by Hastings to translate the Muslim legal commentary, the *Hedaya*. Elizabeth Hamilton joined him in London and helped him with the project for several years, and she socialized at this time with her brother's friends, most of them returned Company officers or fellows of the London Asiatic Society. Charles died of consumption in 1792, shortly before he was scheduled to return to India. *Hindoo Rajah*, Hamilton's first novel, was published four years later, and is both elegy and apologia for her brother. Captain Percy, who dies a good if rather long death at the novel's opening, is clearly meant as a portrait of Charles, and Hamilton depicts herself as his grief-stricken sister, Charlotte. But the personal is highly and very precisely political here; the novel is dedicated not just to Charles but to Hastings, who in 1796 had just been acquitted, after an eight-year impeachment trial, of a vast array of charges relating to his service in India. Elizabeth Hamilton's complex representation of the contact between British and Hindu cultures is driven in no small measure by her impulse to protect her brother's memory from years of what she sees as slander against the Company and all who worked for it.

What little scholarship there is on Hamilton's fiction focusses, like most scholarship on fiction of the 1790s, on domestic political concerns, specifically the intellectual politics of Jacobin versus anti-Jacobin and the related gender politics of the response to Mary Wollstonecraft's *Vindication of the Rights of Woman* (1792). The critical consensus is that in these matters Hamilton is a moderate conservative, clearly anti-Jacobin in her resistance to the ideas of William Godwin, but none the less sympathetic to Wollstonecraft's position on women's education.[5] I would like to

---

[5] Felicity Nussbaum, focussing on Maandaara's reaction to the forwardness of English women, finds the novel deeply conservative on women's issues: "*All* women are the object of satire in *The Letters of a Hindoo Rajah*, and the international observer, though extreme in his criticisms, registers complaints familiar from misogynist satire, as a convincing

consider a different set of issues, namely the colonial and ethnographic concerns of *Hindoo Rajah*, focussing on how scenes of science function in the novel, not just in defining what Hamilton sees as the specific characters of India and Britain, but also in elucidating problems of the "scientific" ethnography she practices.[6] I want to explore how *Hindoo Rajah* participates in the

commentator on English femininity"—*Torrid Zones: Maternity, Sexuality, and Empire in Eighteenth-Century English Narratives* (Baltimore: Johns Hopkins University Press, 1995), 172. Other critics have, however, suggested that Hamilton's political views are far more moderate than those of most anti-Jacobin writers and that she was committed to balanced social reform: see, for example, Perkins and Russell's introduction to their edition of *Hindoo Rajah*, 14–18; Claire Grogan's introduction to her edition of Hamilton's *Memoirs of Modern Philosophers* (Peterborough, Ontario: Broadview Press, 2000), 12; Eleanor Ty's *Unsex'd Revolutionaries: Five Women Novelists of the 1790s* (Toronto: University of Toronto Press, 1993), 26; and Janice Thaddeus's "Elizabeth Hamilton's Domestic Politics," *Studies in Eighteenth-Century Culture* 23 (1994): 266. When she wrote *Modern Philosophers* (1800), her second novel, Hamilton made it clear, even while she dramatized the disaster of a life lived in accordance with Godwin's ideas, that she none the less valued his *Political Justice* (1793) as an "ingenious, and in many parts admirable, performance"—Hamilton, *Modern Philosophers*, 36. And in a letter of 1802, Hamilton concedes that she has come around to Wollstonecraft's position: "I used to combat this [women's superiority] with dear Miss W——— stoutly; but experience has taken her side of the question, and the more I see and know of the world, I am the more convinced, that whenever our sex step over the pale of folly...they ascend the steeps of wisdom and virtue more readily than the other. They are less encumbered by the load of selfishness; and, if they carry enough ballast to prevent being blown into the gulf of *sentiment*, they mount much higher than their stronger associates"—quoted in Elizabeth Benger, *Memoirs of the Late Mrs. Elizabeth Hamilton with a Selection from her Correspondence* (London: Longman, 1818), I: 148–49. One critic who does attend to imperial issues in *Hindoo Rajah* is Balachandra Rajan. Comparing it to other early novels about India by women, Rajan celebrates Hamilton's positive engagement with a feminized India, despite what he sees as her somewhat naive support for Hastings: "Hamilton's unstated intention is to argue that a civilization built on feminine principles can succeed in being just and enduring"—Rajan, "Feminizing the Feminine: Early Women Writers on India," in Alan Richardson and Sonia Hofkosh, eds., *Romanticism, Race, and Imperial Culture, 1780–1834* (Bloomington: Indiana University Press, 1996), 155–56. Claire Grogan is, however, troubled by Hamilton's adoption of the authoritative male voice of the scholar of the Orient: "Hamilton cannot represent the East in the language of male Orientalists without implicating herself in the underlying prejudices of their ideological project"—Grogan, "Crossing Genre," 36.
[6] Hamilton had an early encounter with natural philosophy during an excursion from Stirling to Glasgow and Edinburgh around the age of thirteen: "the greatest advantage she derived from this visit, was an introduction to the noted Dr. Moyse, who was then giving a course of lectures on experimental philosophy. The acquaintance thus accidentally commenced, was afterwards cultivated by a literary correspondence, in which the lecturer liberally undertook to direct the studies of his youthful pupil. In after life, it was often a subject of regret to her, that she had not devoted to classical or scientific pursuits the time

complex processes by which science is made to serve the ends of empire in the Enlightenment, and particularly how Hamilton models an exemplary masculinity in the figure of the scientist, who is offered as a prime mover of a new, benign imperial era. For *Hindoo Rajah* provides a compelling example of what Mary Louise Pratt has identified as the rhetoric of "anti-conquest": its ethnographic narrative captures the other in an apparently disinterested scientific gaze, satisfying a European desire to take "possession without subjugation and violence."[7]

II

Consider again, then, Maandaara's account of the parlor science he witnessed in Agra, ostensibly a scene of benign cultural exchange gone terribly wrong. He first receives a shock from an electrical machine of the portable kind used both for medical treatment and to display the range of electrical effects that had been discovered over the past several decades **[Figure 1]**. He is then shown scenes thrown on the wall from a magic lantern: a popular entertainment, but also a tool for teaching the behavior of lenses. Joseph Priestley praises the magic lantern as a diversion not just "to children, and persons unacquainted with the principles of opticks,"

---

unprofitably wasted in music"—Benger, *Memoirs and Correspondence*, I: 46. "Dr. Moyse" is undoubtedly Henry Moyes, a highly successful itinerant public lecturer in chemistry, natural history, and galvanism. Blinded by smallpox as a child, Moyes worked with a sighted demonstrator. His lecture tours throughout Britain were well-subscribed, as was his tour of America. See John A. Harrison, "Blind Henry Moyes, 'An Excellent Lecturer in Philosophy,'" *Annals of Science* 13 (1957): 109–25, and John A. Cable, "The Early History of Scottish Popular Science," *Studies in Adult Education* 4 (1972): 40–41. Hamilton's assumption that natural philosophy was an entirely appropriate activity for women was widely shared by women writers of all political stripes; Wollstonecraft, for one, proposes "experimental philosophy" as a valuable mental exercise for women—*Political Writings*, Janet Todd, ed. (Harmondsworth: Penguin, 1993), 148. On the central role women played in botanical studies in the late eighteenth century, see Ann B. Shteir, *Cultivating Women, Cultivating Science* (Baltimore: Johns Hopkins University Press, 1996).
[7] Mary Louise Pratt, *Imperial Eyes: Travel Writing and Transculturation* (London: Routledge, 1992), 57.

but "even to philosophers themselves, in an hour of relaxation."[8] It is not clear from Maandaara's letter whether his English host offered any instruction with these rather alarming phenomena; certainly Maandaara has no grasp of the mechanical models by which western science confidently explains these effects. Presumably we are meant to read this scene as evidence of the natural superstition of the Eastern mind, which immediately accounts anything unexpected in nature as the work of devils.[9] Moreover, in both instances, it is the distinctively modern and Western scientific machine that is at the root of Maandaara's bafflement; he displays no curiosity about these instruments, apparently unable to attribute to them the strange phenomena he has witnessed.

Throughout *Hindoo Rajah* Hamilton suggests that scientific method and the scientific instrument are signal markers of cultural difference between East and West, a function of her consistent reading of India versus Britain along the template of ancient versus modern. Following Sir William Jones and other members of the Asiatic Societies of Calcutta and London, Hamilton promotes India as offering rich prospects for antiquarian inquiry, as presenting a truly classical culture, bearing still some of the vestiges of a golden age.[10] While such nostalgia for lost origins is a hallmark of

---

[8] Joseph Priestley, *The History and Present State of Discoveries Relating to Vision, Light and Colours* (London: J. Johnson, 1772), 123. In 1785, Henry Moyes had advised Benjamin Rush, Professor of Chemistry at the University of Pennsylvania, on necessary experimental machines for teaching natural philosophy, including in his list "an Electrical Pump with Battery, Luminous" and a "Magic Lanthorn"—Harrison, "Blind Henry Moyes," 121.

[9] Although Hamilton would disagree with Macaulay on the superiority of Indian poetry, she shares his sense that the Indian learning is deficient in its record of matters of fact. Macaulay wrote in his "Minute of 2 February 1835 on Indian Education" that he "never met with any Orientalist who ventured to maintain that the Arabic and Sanscrit poetry could be compared to that of the great European nations. But when we pass from works of imagination to works in which facts are recorded, and general principles are investigated, the superiority of the Europeans becomes absolutely immeasurable"—*Macaulay, Prose and Poetry*, ed. G. M.Young (Cambridge: Harvard University Press, 1957), 722.

[10] Sara Suleri remarks that while Indian ancientness inspired respect in Burke and Hastings, it evoked contempt in James Mill and Thomas Babington Macaulay—*The Rhetoric of English India* (Chicago: University of Chicago Press, 1992), 33. On the dangers of the residue of such nostalgia for origins in post-colonial criticism, see Spivak, "Can the Subaltern Speak?" 281.

imperial writing, Hamilton makes it strikingly explicit and carefully embeds it in the history and topography of India. She repeatedly stresses the extreme antiquity of the Hindus, whose "annals trace [their origins] back to a period so remote, so far beyond the date of European Chronology, as to be rejected by European pride" (57). And the landscapes of India she paints in the first part of the novel provide vistas of glorious ruin rendered with the elegiac grandeur typical of travel books to Greece and Italy:

> The Pagodas, whose lofty summits had sustained the clouds, and palaces which had once spread their golden fronts to the sun, proud of being the residence of the ancient Rajahs of our nation, now bow their time-worn heads to listen to the voice of strangers, and behold the sacred characters, inscribed on their bosoms, familiarly perused by a people, whose nation had not sprung into existence at the time these towering monuments of Eastern splendour had commenced the progress of decay. (152)

The exciting difference for Hamilton is that India offers to the modern Western antiquarian the opportunity to witness *living* antiquity; while it is now impossible to trace any of the heroism of the past in the faces of the modern inhabitants of Greece and Italy, "in Hindoostan, the original features that marked the character of their nation, from time immemorial, are still too visible to be mistaken or overlooked" (71), a consequence, Hamilton believes, of the rigidity of their religion and caste system. For the first time, the ancient psyche is available for study, and in her portrait of the pious and feeling Zaarmilla, Hamilton imagines what this psyche might be. But for Hamilton Hindu culture is also clearly classical in that it is a profoundly literary culture. Like the ancient Greeks or the tribes of Israel, the Hindus live lives governed by texts that articulate their national and religious desti-

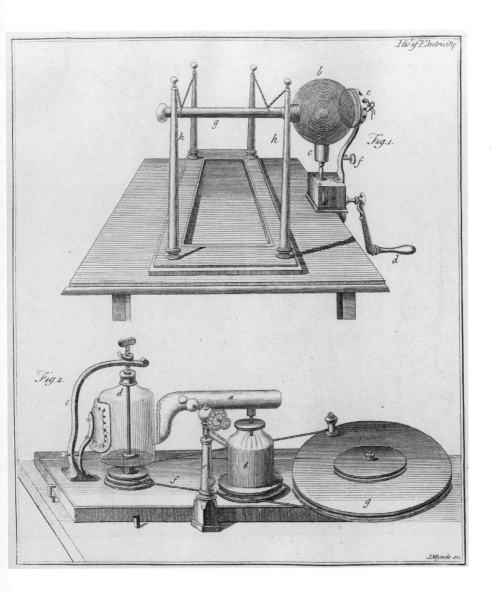

Figure 1: Illustration of portable electrical machines, from Joseph Priestley,
The History and Present State of Electricity (1775), II, plate 1.
Courtesy of the William Ready Division of Archives and Research Collections,
Mills Memorial Library, McMaster University.

Figure 2: Illustration of the Observatory at Benares, from Robert Barker,
"An Account of the Bramin's Observatory at Benares,"
Philosophical Transactions 67 (1777), plate 12.
Courtesy of the Gerstein Science Information Center,
University of Toronto.

nies. Hamilton has her characters quote snatches of epic poetry throughout, rather as if they were quoting Hesiod or Virgil. And she promotes the idea that the great Indian epics—now handily available to the English reader through the elegant translations of Sir William Jones—are the equals of, indeed possibly the sources of, the epics of Greece and Rome. In the "Preliminary Dissertation" on Hindu culture prefixed to the novel, Hamilton stresses the uncanny consonances between the Hindu and Greco-Roman pantheons (64). In fact, the satire of British culture that forms the bulk of the novel is built upon this very ancient investment in the literary; Zaarmilla has read the gospel—what he calls the "English shaster"—which had been given to him by the dying Captain Percy. To often hilarious mock-heroic effect, in England he tries to read British culture against what should be its governing text and comes up baffled every time. He is horrified at the prospect of a rudderless society, living without reference to the Word.

In thus rendering Hindu versus Briton as a matter of as ancient versus modern, Hamilton depicts the clash of two profoundly antithetical epistemologies: the one literary and poetic, the other empirical and scientific. This is nowhere more evident than in another scene of science in the novel, this time of ancient Hindu science forgotten. European orientalists granted that the Hindus did excel in one branch of the philosophy of nature, astronomy. In 1777, Sir Robert Barker described his visit to the observatory at Benares in the *Philosophical Transactions*, an account accompanied by gorgeous illustrations of the massive stone machines **[Figure 2]**, some of which are clearly still in use by the Brahmins, and which Barker guesses are at least 200 years old.[11] In *Hindoo Rajah*, Hamilton offers a fictional rendering of Barker's visit. Zaarmilla has made friends with Percy's fellow officers, and in the company of two of them he visits the Benares observatory:

[11] Robert Barker, "An Account of the Bramin's Observatory at Benares," *Philosophical Transactions of the Royal Society of London* 67 (1777): 598–607.

> Both these gentlemen were deeply learned in this divine
> science [astrology]. The stupendous engines, constructed
> by the ingenuity of our ancestors for measuring the
> expanse of heaven, and tracing through its tractless arch
> the path of its illustrious inhabitants, filled their minds
> with astonishment. Alas! that these evidences of the
> wisdom of fathers should now serve to mark the degener-
> acy of their children!...These strangers could, at one
> glance, comprehend the use of those instruments, which
> the Pundits, who attended us, could not explain. (154)

Here again, Hamilton shows us the Hindu baffled by the machine,
which can only be illuminated by the glance of the Western
scientist. Zaarmilla seems strangely untroubled by this moment
of cultural humiliation. He likens the Goddess of Science to the
sun: "when we vainly imagine she is gone for ever, if we open the
eyes of our understanding, we shall see her beaming with redoubled lustre on the children of another hemisphere" (155). This
scene fits, of course, within Hamilton's larger justification of
Britain's place in India. Only through the penetrating gaze of the
Englishman can India become legible, not just to world but to
itself. But this is also a scene about cultural destinies. Hamilton
needs to show us that the one claim Hindu culture was once able
to make to the philosophy of nature is now defunct. The sun of
science smiles on the land of Newton alone, and clearly for
Hamilton Britain's future as an imperial power depends on the
labor of the scientist and his mastery of the machine.

III

When Zaarmilla finally does travel to Britain, he witnesses a broad
cross-section of life, and his naive record of the vice and eccentricity of the British at home forms the latter half of the novel.
Hamilton's satiric inversion of the imperial gaze is, as intended,
thoroughly unnerving, and the broad scope of her sat-

ire—encompassing politics, religion, learning and the arts—dizzy-ing. In the crowded scene, however, one British character does receive a more sustained consideration: Dr. Severan, a gentleman of modest but independent means. Zaarmilla brings letters of introduction to Severan from one of Percy's fellow officers, and Severan takes the increasingly muddled Rajah under his wing, doing his best to explain some of the oddities of British life. Severan, who in his piety, charity and industry comes to represent all that is good in the British character, has devoted his life to chemistry. Zaarmilla first encounters him surrounded "by numerous odd shaped vessels, some of glass, others of metal, but for what use I could not possibly comprehend" (208). On a subsequent visit, distracted by his conversation with Zaarmilla, Severan forgets to attend to a retort he is heating, and the apparatus explodes, filling the air with "suffocating effluvia" that almost drive Zaarmilla from the room. This catastrophe, which destroys the work of many weeks of experiments and sends Severan into a momentary despair, shortly proves to be epiphanic: Severan, "who never lifted his eyes, from the remains of the broken vessel, suddenly clapping his hands together, exclaimed, in a transport of ecstacy, 'I see it! I see it!—Heavens! What a discov-ery!'" (223). The appearance of the matter remaining in the retort suggests a radically new course of experimentation which prom-ises, he tells Zaarmilla, "a most important discovery." This work at the very vanguard of science requires an additional apparatus costing fifty pounds, an amount Severan cannot afford, even with the most severe retrenchments, and an amount he will not accept from Zaarmilla, arguing that such indebtedness would ruin their friendship. But just as he has resigned himself to the will of God, with providential felicity the exact sum unexpectedly comes Severan's way. In an instant the philosopher is dreaming of commencing work, but his dreams vanish when the distressed wife of a friend bursts into the room, explaining that her husband has just been imprisoned for a debt of fifty pounds:

"Fifty pounds!" repeated the philosopher.   "And fifty
pounds would release your husband from a jail. Fifty
pounds would restore a father to his infant family, and
make the heart of a virtuous woman rejoice. It is the
noblest of all experiments!" (239)

The obligatory excursion to debtors' prison, rich in pathos,
follows. Reading the pious and improving reflections on charity
that ensue, one might be tempted to think that Dr. Severan's
laboratory functioned as little more than an exotic setting to
enliven otherwise fully predictable scenes of sentiment, scenes
that are the stock-in-trade of fiction of this period. I would argue,
however, that in her portrait of Severan—the exemplary British
natural philosopher and, in many ways, a masculine ideal—Hamilton argues for science's central place in the destiny of Britain and
its people.[12]

I should register first, given the ethnographic sweep of *Hindoo
Rajah* and its program of mapping British culture, Severan's role in
Hamilton's agenda of constructing a British national identity, her
proposal that the Christian natural philosopher is the intellectual
and moral center of British society. Of course the myth that
natural philosophy was somehow the peculiar property of Britain
was still alive and well in Hamilton's day. Joseph Priestley, for
one, was always happy, in his popular histories of the sciences, to
depict natural philosophy as a legacy lovingly passed from Bacon
to Newton to Priestley himself: he even fantasizes in print about
bringing Newton back to life and entertaining him with all the
electrical discoveries made since his death.[13] In *Hindoo Rajah*, as
we have seen, Hamilton has made the scientific machine a central
trope in her cross-readings of Indian and British culture. Severan,

---

[12] Hamilton works in direct contrast with Montesquieu, whose Persians in Paris could only
deplore the near-madness of the European scientists they encountered, the frenzied
alchemist or the measurement-obsessed geometrician—Montesquieu, *The Persian Letters*,
John Ozell, trans. (London: J. Tonson, 1722), I: 147–50 and II: 189–94.

[13] Joseph Priestley, *The History and Present State of Electricity*, 3rd ed., 2 vols. (London: J.
Johnson et al., 1775), II: xv.

British scientist, is clearly the master of his enormously complicated, if occasionally temperamental machine, and in the end, of course, Hamilton does let Severan buy his new apparatus and continue his research. We are clearly meant to read Severan as a figure of utter modernity, and to view his work as culturally transformative, fashioning a brighter future.

That said, what is striking about Hamilton's portrayal of Severan is her emphasis on the personal, on the ethical and affective dimensions of the scientific life, and again there are echoes of Priestley. Zaarmilla, as he watched Severan at his research, was thoroughly impressed by his serious diligence and application, and particularly remarked on "the glow of pleasure which animated his finely expressive countenance" as he pursued nature (236, see also 305). Hamilton, who carefully distinguishes these mental passions from sensual pleasures (236), makes a case, here, for the personal rewards that attend the life of science: that the *vita activa* of experimental inquiry leads to a happier, more virtuous existence, and that God has attached special and higher pleasures to inquiries into the workings of creation (223–4).[14] This theme is prominent in the work of Priestley, who had argued that "human happiness depends chiefly upon having some object to pursue, and upon the vigour with which our faculties are exerted in the pursuit"[15]—a discipline, he points out, readily provided by a course of scientific enquiry. But Hamilton, a lifelong and devoted adherent of the Church of Scotland, also joins Priestley in wanting to bring natural philosophy into tune with the religious awakening of her generation, and to effect a Christian reform of British culture.[16] Priestley emphasized the necessary humility of

---

[14] This theme resurfaces in *Modern Philosophers*, where natural philosophy is praised as having "a direct tendency to influence the moral character of man"—Hamilton, *Modern Philosophers*, 311.

[15] Priestley, *History of Electricity*, II: xiv.

[16] As Nigel Leask has observed, "Hamilton's patrician evangelism is more concerned, in the troubled climate of 1796, to 'christianize' Britain than India"— *British Romantic Writers and the East* (Cambridge: Cambridge University Press, 1993), 101. Hamilton came from a family with a strong tradition of dissent, and the aunt who raised her in Stirling was a firm Presbyterian. When in London Hamilton would conform to the rites of the Church of

the natural philosopher in contemplating God's work, arguing that the greatest discoveries in science have been accidental rather than projected, the true philosopher's mind ever subservient to the wondrous workings of the natural world. Certainly Severan's work proceeds with such humility, indeed seems peculiarly a direct and immediate engagement with the divine will. We are meant to sense God's hand, most loving at the moment it seems most cruel, at work in Severan's exploding retort and in the peregrinations of his fifty-pound note. Moreover, Priestley had argued that however useful natural philosophy may prove in the practical amelioration of the conditions of life, by far its chief benefit is to the condition of the soul of the scientist: "the greatest, and noblest use of philosophical speculation is the discipline of the heart, and the opportunity it affords of inculcating benevolent and pious sentiments upon the mind." The true natural philosopher is conscious always of final causes, even as he busies himself about efficient ones; as he witnesses the order and beauty of creation, his heart is readily filled with "unbounded love, gratitude and joy."[17] All these ideals are clearly embodied in Hamilton's Dr. Severan, who is the one Briton Zaarmilla has found who seems to live in accord with the doctrine of love inscribed in the English shaster. It is not just that Severan can choose charity over knowledge, but that God's work and the work of science are one: to liberate his friend from prison is, in Severan's words, "the noblest of all experiments."[18]

England, but she worried that her friends might think she had changed her faith, and that exposure to the English rite might cause in her a "greater laxity in spirit"—quoted in Benger, *Memoirs and Correspondence*, I: 263–65.

[17] Priestley, *History of Electricity*, II: xxii–xxiii.

[18] This theme of the civilizing effects of natural philosophy is evident in Hamilton's promotion, elsewhere in the novel, of science as a pastime for the gentry. The admirable Mr. Darnley is a squire who has given up hunting for natural history, one for whom "the study of Mineralogy and Botany, and exquisite relish for the beauties of nature, refined by an acquaintance with the sister arts of Poetry and Painting, gave sufficient interest to rural scenery, without any aid from the misery of inoffensive animals" (293). Likewise the benign Mr. Sydney in *Modern Philosophers* is a dedicated amateur botanist and entomologist with his own collection of favorite curiosities.

In case we had any doubts about all this, Hamilton offers us, for contrast, the portrait of a mad scientist, of a self-indulgent, undisciplined pretender to science who is, Hamilton makes clear, no scientist at all. Sir Ardent Caprice falls, as his name suggests, within the clearly established type of the aristocratic dabbler whose passions are intense but short-lived. He is, moreover, a free-thinker and a radical, and when he turns to science it is only to prove his darling theories. In the words of Zaarmilla, Sir Ardent Caprice is "a worshipper of systems" for whom "the fair face of Nature has no charms" (263). Taking fire at the proposition of his young friend, Mr. Sceptic, that all behavior, human and animal, is merely a product of instinct, and that these instincts are formed by external circumstances, Sir Ardent Caprice sets out to prove it by what he believes to be an "experiment," aided by his friends Sceptic and Axiom. He has his servants gather all the sparrows on his estate and place them in an oversized bee hive constructed for the purpose, proposing that they should all soon be humming and gathering honey like bees. Finding the next morning that these sparrows have remained unregenerately sparrow-like, in fact have all escaped the hive and gone about their daily business, he orders his servants to gather all the nestlings they can find and place these in the bird hive, rubbing their tiny bills with a little honey for good measure. This too proves a failure: after three days the hive is opened to reveal a

> Sight of horrors! and smell, still worse than the sight! The lifeless corses [sic] of the three hundred half-fledged nestlings lay at the bottom of their hive, in a promiscuous heap. —"They have effectually swarmed at last!" said Mr. Axiom.—Neither the Baronet, nor the young philosopher, staid to make any remark—but every one putting his fingers to his nose—impelled by the *necessity* of *existing circumstances*, hurried from the scene. (269, Hamilton's emphasis)

Clearly Hamilton's satire here is directed at the Jacobins for their dangerous attachment to *"abstract principle"* (307). The violence done to nature by such a deductive approach to science is grotesquely figured in the sight of the massacred nestlings. But in this scene Hamilton is also working hard to reinforce, by antithesis, the true scientific method and its centrality to personal and national virtue. We are clearly meant to compare Severan with Sir Ardent Caprice; the experiments of each come to grief, and in the process produce a powerful stink, but where the former learns from nature the latter simply shifts his attention elsewhere. Moreover, Zaarmilla carefully contrasts the approaches of the two philosophers, noting how Sir Ardent Caprice "disdains the slow process of experiment, and chiefly glories in contradicting common sense. [His] main object is, to shew that the *things which are, are not*, and the *things which are not, are*: and this is called Metaphysics" (248–49). Hamilton thus uses the naive voice of the Hindu traveller to announce the death of metaphysics in Britain— a coffin, you might think, that already has plenty of nails in it, but Hamilton clearly believes that radical thought threatens everything British and does so by virtue of being metaphysical, presumptive, unempirical, unscientific.

Hamilton's empiricism is natural enough for one writing in an anti-Jacobin tradition. April London has shown how the anti-Jacobins, seeing their adversaries committed to utopian social romance, countered with satire and history. Privileging, in Robert Bisset's words, "experience of fact," they strove "to convert naive idealists into sceptical realists."[19] But Hamilton's attachment to the scientific method is also a product of her early reading of Locke, and she recasts his ideas to bring them into tune with the affective and religious priorities of her generation.[20] So when

---

[19] April London, "Novel and History in Anti-Jacobin Satire," *Yearbook of English Studies* 30 (2000): 73. London quotes from Robert Bisset's *Life of Edmund Burke*, 2 vols., 2nd ed. (London: G. Cawthorn, 1800), II: 304.

[20] Locke's epistemology and psychology clearly inform Hamilton's views on the working of the mind and particularly on education, from *Hindoo Rajah*, her first book (see for example 214 and 221), to her final revisions for her *Letters on the Elementary Principles of Education*, 2 vols., 5th ed. (London: J. Johnson et al., 1810).

Hamilton grants Severan the final word on these issues, his speech blends established Lockean epistemology and a new vigorous Protestantism:

> almost all the errors of metaphysicians have arisen from their neglect of natural philosophy.—The extreme accuracy, and exact precision, that is requisite in the investigation of the phenomena of the material world, would induce like habits of reasoning in regard to that of the mental: while the Colossus of Scepticism, I mean Atheism, would, by an acquaintance with the works of nature, be utterly annihilated. (273)

For Hamilton, then, natural philosophy is the queen of the sciences, and the experimental method the engine driving British progress and the basis of Britain's claim to cultural supremacy. And for Hamilton, as for her brother Charles and as for the fellowship of the Asiatic Societies of Calcutta and London, oriental scholarship unquestionably partakes of the method and the glory of British science.[21]   Whether in his circumspect, empirical ethnography or his painstaking decoding of ancient inscriptions, Hamilton's Christian orientalist is, as much as her Christian chemist, an agent in Britain's destined ethical and material conquest of the world.   In this vein Hamilton quotes, in her preface, from Thomas Maurice's *Indian Antiquities* (1793–94): "The buried tablet has been dug from the bowels of the earth; the fallen, and mouldering pillar has been reared; coins, and medals, struck in commemoration of grand, and important events, have been recovered from the sepulchral darkness of two thousand years; and the obsolete characters, engraved on their superficies, have, with immense toil, been deciphered and explained" (66). India in effect

---

[21] Ethnography had, of course, long been considered part of the discipline of natural philosophy: scientific travellers had long treated native peoples as part of the fauna of the lands they explored, and spears and masks jostled with fossils and eggs in Britain's museums.   The Asiatic societies, which modelled themselves on the Royal Society of London, mixed the cultural and the natural indiscriminately.

becomes nature here, only revealing herself to the patient industry of the scientific orientalist, who in turn is dedicated only to bringing her mysteries to light.

IV

Hamilton's strenuous promotion of both science and orientalism draws upon larger and noisier debates in her culture, namely those of masculinity and sensibility, staple themes in novels of her period. Like Frances Burney and Jane Austen, she takes a carefully balanced position on masculine sensibility, warning against its excesses. Thus in *Hindoo Rajah*, we are asked to take stock of the character of Delomond, a man of feeling who, upon meeting with some disappointments in his attempts to build a fortune in India, determines to retire in poverty to England rather than persevere: "it is not a little to be regretted, that this amiable man frequently indulges a certain soreness of mind, which may not improperly be termed the illegitimate offspring of sensibility. What proves its spurious birth, is, that while genuine sensibility is ever alive to the feelings of others, this bastard branch of the family, is only mindful of its own" (197). Such debilitating affective refinement in a man is as selfish and as dangerous, Hamilton implies, as the aristocratic puppyism of a Sir Ardent Caprice. Her heroes, Dr. Severan in Britain and Zaarmilla's friends in the East India Company, Percy, Grey and Denbigh, all model a masculinity in which sensibility is tempered by Christianity and personal industry. Sensibility becomes inextricable, here, from the appropriate emotions of charity, friendship, and family, all displayed in many scenes of sentiment, as when Captain Grey comes to the aid of Mr. Morton (171–74), the destitute son of the curate who was his first teacher, or when Mr. Denbigh returns from India to the arms of his aged parents (285–86). Hamilton's account of natural philosophy as balancing the affective and the ratiocinative, as marrying emotions of awe at the glory of creation with a diligent and circumspect empiricism, is perfectly in tune with the disciplined masculine sensibility that was the goal of so much fiction at the century's end. Mary Louise Pratt, in her reading of Mungo Park's highly affective *Travels into the Interior Districts of Africa* (1799), argues that sentiment and science work together in travel writing of this period to create a "drama of reciprocity" between the questing European and the newly discovered other.[22] In much the same way as Park, Hamilton chooses the affective realm to claim the affinity of Hindu and Briton, projecting a refined sensibility onto both her Hindu characters and her East India Company officers. Zaarmilla makes his first appearance as a figure of charity; he shelters two sick and wounded Rohilla refugees, despite the fact that the Rohillas have been until now the oppressors of Zaarmilla's people. And it is, of course, the New Testament's message of love that strikes a cord in Zaarmilla's heart and convinces him of its truth. Finally, Hamilton emphasizes the integrity and profundity of the friendship that binds Zaarmilla to Severan, Percy, Grey and Denbigh. When she moves to characterize the Hindus as a whole, she describes them as a "mild and gentle race" (62), a people of an "amiable and benevolent character" (57). What the Hindu lacks, Hamilton makes clear, is the intellectual discipline of the Briton, the industry and active engagement in the present displayed not just by Severan the scientist but the novel's orientalists as well. Captain Percy himself, before he appeared on Zaarmilla's doorstep wounded, had been ranging northern Hindustan in search of antiquities (81). Zaarmilla, at best, is meant to be read as the superior of his friend Maandaara, who exemplifies the intellectual apathy and superstition typical in caricatures of the East. Zaarmilla is something of a hybrid who holds out, in his intellectual curiosity, the possibility of a bridge between cultures. But both Hindus, as figures of feeling without improving employment, verge dangerously on the feminine. Even as the novel's ethnographer of Britain, Zaarmilla functions more as observer than agent, and even in this he depends heavily on the guidance of Severan to set him straight.

---

[22] Pratt, *Imperial Eyes*, 81.

Hamilton's intriguing fusion of the discourses of science, masculinity, and sentiment is made to serve, as I have argued, her program of national social renewal. Even as her satire exposes the nation's ills, she provides, in the figure of the scientist-orientalist, a savior not just for the nation but for the world. Hamilton argues implicitly that, just as England needs the virtue and labor of the Christian natural philosopher, India needs the energy and diligence of the East India Company officers to rid it of the curses of tyranny and idleness. Zaarmilla is the first of a new generation of Indians, inspired by the English shaster, ennobled by English friendship, and fired by the example of English industry. But if Hamilton's moral panorama is global in scope, it centres on a figure very close to home. Charles Hamilton, who like his sister had begun life with slender financial prospects, had served Hastings and the Company diligently with his sword and his pen, and was well rewarded with advancement and well established within the colonial governing elite. He died just before returning to a new position as British Resident to the court of the Nawab of Oudh, a position that would have provided ample opportunity for personal enrichment. Charles was, in short, well on his way to becoming a nabob. In fact, in a letter to Charles of 1783, Hamilton playfully concedes that she has long imagined him as such: "I beheld you placed on a lordly seat, with all the grandeur I have ever seen displayed by any of our eastern nabobs."[23] The figure of the returning merchant-adventurer, rich beyond measure, was well established in the British imagination. But the caricature of the nabob was given a particularly vivid and ugly reincarnation by Burke, whose rhetoric sought to provide a focus for moral outrage in the persons of Hastings and his officers. As Sara Suleri has shown, Burke consistently depicted the Company officers as greedy and self-indulgent boys, who lived in utter ignorance of the lives and customs of the people they governed and who were incapable of the profound cross-cultural sympathy which Burke

---

[23] Quoted in Benger, *Memoirs and Correspondence*, I: 94.

himself so amply displayed.[24] It is, of course, this very sympathy that Hamilton works so hard to reclaim for those serving in India. But Burke also turned his rhetoric against the figure of the rich nabob returning home:

> Arrived in England, the destroyers of the nobility and gentry of a whole kingdom will find the best company in this nation at the board of elegance and hospitality. Here the manufacturer and husbandman will bless the just and punctual hand that in India has torn the cloth from the loom, or wrested the scanty portion of rice and salt from the peasant of Bengal, or wrung from him the very opium in which he forgot his oppressions and his oppressor. They marry into your families; they enter into your senate; they ease your estates by loans; they raise their value by demand; they cherish and protect your relations which lie heavy on your patronage.[25]

Far from being a welcome addition to the national wealth, the nabobs' money, cruelly extracted, is poisoned at the source. Brought home, their insidious trade in power and prestige becomes a cancer in the British body politic. By contrast, Hamilton shows her Company officers as selfless servants of both the Crown and the peoples they govern, paying with their healths and even their lives for the promise of a new India. While, in her struggle to make India accessible to her readers, Hamilton may at points inadvertently domesticate the Hindu, her first agenda is that of redomesticating the returning nabob.

---

[24] Suleri, *Rhetoric of English India*, 32–33.
[25] "Speech on Mr. Fox's East India Bill (December 1, 1783)," in *The Writings and Speeches of the Right Honourable Edmund Burke*, 12 vols., ed. Paul Langford (Boston: Little, Brown, 1901), II: 464.

V

But there is one final scientific machine in *Hindoo Rajah*. At his death, Captain Percy leaves Zaarmilla his Bible, his sister's portrait along with some of her poetry, and his repeating watch. Altogether, the gifts hold out the promise of alternative colonial relations, relations informed by love and respect, rather than power and profit. Individually, the gifts suggest some of the specific pathways such a new economy might take. The bible, of course, symbolizes for Hamilton the greatest gift from West to East, that of salvation. Charlotte Percy's picture and poetry have more complicated resonances. They speak of the affective bond between Hindu and Briton, the brotherhood the novel seems to promise at its outset. Percy's "gift" of this sister is replicated soon after in the novel, when Zaarmilla and Maandaara exchange sisters as wives in token of their abiding affection for one another. But at the same time the fictional Charlotte Percy is clearly to be read as the real Elizabeth Hamilton. By thus placing herself and her writings in Hindu hands, Hamilton seems to want to draw attention to her own exemplary trust and her commitment to bridging the gulf between Britain and India.[26] All these gifts are treasured by Zaarmilla, but Percy's repeating watch—the gift of western technology and, by the by, of western industry and discipline—he treats with more ambivalence, calling it a "little shrill-voiced monitor, whose golden tongue proclaims the lapse of time" (95). Time, it would seem, simply does not matter to an ancient.

The oddity of Zaarmilla's whimsical treatment of the watch, his sudden circumspection jarring with the high-flown sensibility

---

[26] Writing to a friend about her doubts about *Hindoo Rajah*, still in manuscript, Hamilton refers to it as "my black baby," a signal of her extreme diffidence, but one that hints too at the fantasies of cultural crossings at the heart of the novel—quoted in Benger, *Memoirs and Correspondence*, I: 126. See Claire Grogan's important reading of Hamilton's transgressive image for her hybrid test: "Hamilton undermines the existentialist opposition of color (black/white), sex (male/female), and genre (fiction/non-fiction, novel/Oriental study) by producing, as a white woman, a black child"—Grogan, "Crossing Genre," 37–38.

of his mourning for Percy, hints at Hamilton's struggles to contain the disparate energies of her fiction. This seems to be one of those places in the novel, and there are many, where her chosen genres of satiric travel book and sentimental novel prove to be at odds with one another, and where her political agenda of vindicating the East India Company is out of sorts with her concurrent attack on the materialism and self-indulgence of her generation. Moreover, this seems a place where Hamilton's Enlightenment conviction in science's power to regenerate her culture (so troublesome for readers in a post-technological and post-colonial age) falters for a moment. Percy's repeating watch resonates with all those other wondrous and expensive machines in *Hindoo Rajah*, but casts them in a new and troublesome light. It conjures up exploding retorts and bird-hives full of dead nestlings, and recalls the electrical generator and magic lantern owned by that English Saib in Agra, who first gave his Hindu guests electric shocks so severe they felt their arms were broken, and then horrified them with scenes of carnage, of marching armies, charging elephants, drowning sailors. The implicit link here between Western technology and imperial violence evokes all those thoroughly modern machines of war that were enforcing Britain's economic will in the subcontinent. Science here is about violence and domination. This is, of course, awkward, in that *Hindoo Rajah* sustains a wilful aphasia when it comes to the violence of British rule in India, dumb about the torture and extortion and warmongering that Burke so thoroughly chronicled.[27] Only once or twice does Hamilton's staunch silence on these issues give way, and then only to insistent, almost

---

[27] By contrast, in *Modern Philosophers*, where India and by association her brother's reputation are not in question, war and the vices of an often idle military do become a subject for Hamilton's satire, and on this theme she seems very close to Wollstonecraft: "The wretched remains of those numerous armies which in the beginning of the contest had marched forth, elate with health and vigour, were now returned to their respective countries; some to languish out their lives in hospitals, in the agony of wounds that were pronounced incurable; some to a wretched dependence on the bounty of their families, or the alms of strangers; and the few whose good fortune it was to escape unhurt, according to the seniority of their regiments, either disbanded to spread habits of idleness and profligacy among their fellow-citizens, or sent into country quarters to be fattened for fields of future glory"—Hamilton, *Modern Philosophers*, 77.

panicky denials of Burke's claims of the greed and brutality of Hastings' governorship, that he and his officers made themselves "splendid by the beggary and massacre of their fellow-creatures" (67). Such denials are fuelled, of course, by her eagerness to vindicate her brother from such charges. Zaarmilla himself describes the gallant behavior of the victorious British at the battle of Cutterah (where Charles Hamilton fought), saying that the British soldiers did all they could to protect the vanquished Rohillas from the vengeance of the Hindus. And all the Hindus in the novel cast Hastings as a savior and liberator (108). When Hamilton does concede the violence of British rule, she tries to consign it to the past, implying that a new, scientific spirit is at work in the colonial government that will atone for the sins of the past:

> The thirst of conquest and the desire of gain, which first drew the attention of the most powerful, and enlightened nations of Europe toward the fruitful regions of Hindoostan, have been the means of opening sources of knowledge and information to the learned, and the curious, and have added to the stock of the literary world, treasures, which if not so substantial, are of a nature more permanent than those which have enriched the commercial. (55)

A new benign form of conquest and a new immaterial commerce will replace the old, but Hamilton's phrasing here seems to concede that the new ethnography has roots in the old colonialism. Her elegant East India Officers, however much they pose as gentle antiquaries, are still in uniform.

And finally, for all her commitment to the promise of a new empire of the mind where the East can be diligently mined for the treasures of knowledge, and for all her conviction of the benefits to India of being pulled into modernity, Hamilton seems to have written a book that proves, again and again, the difficulty of knowing the other. The long process of Zaarmilla's disillusionment with Britain is marked by his repeated failure, despite his best intentions, to make sense of what he sees. Where Britain is concerned, Hamilton slips into parodic ethnography, with strong echoes of Swift's *Travels* (1726). Swift, too, had mocked Britons' obsession with measuring time, as when the Lilliputians decide that Gulliver's pocket watch must be his god, since he does nothing without consulting it. When Captain Percy gives an initial glowing account of Britain to Zaarmilla, he sounds uncannily like Gulliver before the King of Brobdingnag, his reason swallowed up by his love of nation. Likewise Zaarmilla in England seems like Gulliver in Houyhnhnmland, too eager to build a place of liberty and reason. However much, under the guidance of his British hosts, he tries to clear his eyes of the accretions of culture—and he is very self-conscious in striving to render "a faithful copy of the first impressions made upon [his] mind" (162)—he almost always gets it wrong, mistaking, for example, English ladies at a rout for dancing girls (168). Most of all, Zaarmilla cannot free himself from the urge to give predominance to the literary, to read Britons against the Gospel, assuming it must be the "guide of their practice," and to try, valiantly, to make life and text connect. His final resort, in this predicament, is to deduce that there must exist a second, secret Gospel governing British behavior, which "must very essentially differ from the old one" (240) and which would explain the many absurdities he has witnessed. In the end, the intensity of Hamilton's anxiety about the secularization of Britain raises inevitable questions about Britons' ethical fitness to rule the world.

Zaarmilla's many mistakings suggest that Hamilton herself is far from comfortable with her role as ethnographer, obsessively conscious as she is of the slipperiness of the ethnographic fact. In attacking Burke, Hamilton consistently claims that he simply does not know India (147n); he is, she implies, a parliamentary Sir Ardent Caprice, his projections about the Hindus no different from the mad scientist's attempts to transform sparrows into bees. Burke's blunders are, however, symptomatic of a national disease. Empiricism is under siege in *Hindoo Rajah*, which portrays a Britain in which fictions everywhere parade as fact. Zaarmilla is baffled by the novels and romances he has been given, which claim to be "histories" and "authentic memoirs" but bear no resemblance to human life (189–92), and he is horrified to learn that British newspapers are filled, for the most part, with bold-faced lies (244). The disease is, of course, most virulent among the Jacobins, where, thanks to Rousseau, romance and metaphysics are indistinguishable.[28]   Interestingly, in *Modern Philosophers*, her next novel, Hamilton will make ethnography itself the shoddy vehicle of Jacobin delusion. Here she depicts a circle of Godwinian zealots who have read François LeVailliant's *Travels into the Interior Parts of Africa* (1790) and, enchanted at the prospect of a life without government and social distinctions, determine immediately to go and live among the Hottentots.[29] So, however much Hamilton trumpets the power and rectitude of empiricism, she everywhere shows its failure in Britain, an anxiety that reflects finally on her own tenuous position as ethnographer-novelist. Choosing to battle the Jacobins in their own genre, writing romance against romance, she finds herself actively embarked in the national trade in fiction passed off as fact. And when it comes to India, she is just as vulnerable as Burke to the objection that she has never witnessed the East first-hand. Like Zaarmilla, she finds herself relying on texts—on travel books and ethnographies—to understand the other. No wonder, then, that *Hindoo Rajah*, which began with the promise of making the Hindu familiar to us, like a brother, must end with the Hindu turning away in confusion and dismay, a sparrow who will not be made into a bee.

---

[28] Hamilton, *Modern Philosophers*, 38 and 86.
[29] Hamilton, *Modern Philosophers*, 141.

A GOD WHO MUST
Science and Theological Imagination

Paul K. Johnston
State University of New York at Plattsburgh

*God don't never change. He's God, always will be
God.* — Blind Willie Johnson

I t has become a commonplace in thinking about science, religion, and culture to assert that, one way or another, science and conservative religion are at odds with one another: that either the rise of science has brought about the decline of conservative religious thought and belief, or that conservative religion has asserted itself as a reaction against the dominance of scientific thought and belief, or that, most likely, some combination of these two paradigms has been at work in the three hundred years since the advent of the scientific revolution.[1] In eighteenth-century colonial America, the opposition of progressive scientific thought and religious conservatism might seem obvious in the contrast between two of the period's leading intellectual figures: Benjamin Franklin and Jonathan Edwards.[2] Franklin is remem-

---

[1] For a recent influential statement of this thesis, see Karen Armstrong, *The Battle for God* (New York: Alfred A. Knopf, 2000).
[2] Edwards's religious conservatism seems obvious enough: his sermons and theological treatises are aimed at least in part at combating the liberal tendencies which eventually

bered for many remarkable things, among which are his achievements in science, most famously his experiments and papers on electricity, including the experiment in which he flew a kite with a metal key attached to the near end of the string in a thunderstorm, an experiment that led not only to his more practical invention of the lightning rod, but to scientific papers in which he introduced many of the terms still used to discuss electricity. Edwards, on the other hand, is perhaps most remembered for the sermon he preached at Enfield, Connecticut, in 1741, during the religious revival that came to be known as the Great Awakening. "Sinners in the Hands of an Angry God" seems a manifestation of another world from that which produced Franklin and his kite experiment. Its images—the spider held over the fire, the arrow ready on the string to be made drunk with the blood of the sinner's heart, the sinner walking over the pit of hell as on a rotten covering, the God who will "laugh and mock" in response to cries for pity from the tormented in hell—are not meant to speak to scientific reason, but to evoke emotion. "Though [God] will know that you cannot bear the weight of omnipotence treading upon you," Edwards told the crowded congregation at Enfield, "yet he will not regard that, but he will crush you under his feet without mercy; he will crush out your blood, and make it fly, and it shall be sprinkled on his garments, so as to stain all his raiment."[3] Contemporary liberal detractors were critical of the emotional excesses such preaching evoked—weeping, wailing, shrieking, and so forth—emotional excesses that seem a far cry from the detached scientific reason presented by Benjamin Franklin.

---

undermined the Calvinism Edwards championed. Yet Alan Heimert, *Religion and the American Mind: From the Great Awakening to the Revolution* (Cambridge: Harvard University Press, 1966), viii, notes at the outset of his study that Edwards's thought was in its own way radically democratic, while holders of more liberal views often tended to social conservatism.

[3] "Sinners in the Hands of an Angry God," *The Sermons of Jonathan Edwards: A Reader*, ed. Wilson H. Kimnach, Kenneth P. Minkema, & Douglas A. Sweeney (New Haven: Yale University Press, 1999), 60. Though the authoritative texts for Edwards's works are those published by Yale in its *Works of Jonathan Edwards* (1957–), the publication of his sermons is not yet complete. In the meantime, Yale has published this collection of fourteen sermons from the approximately 1,250 that have survived.

Yet Edwards was far from untouched by the advent of science in Western culture, nor was he consciously reacting against it, however remote from it he may have seemed at Enfield that day. The earliest writing we have of Edwards, most scholars agree, is a piece of scientific investigation. Once thought to have been written by Edwards when he was as young as eleven, "Of Insects" is now thought more likely to have been written a few years later, when he was a student at Yale, though still only in his teens.[4] The problem Edwards set out to explain was the means by which a certain species of spider is able to spin webs between trees separated by a distance of a hundred feet or more. To solve this problem, Edwards observed the spiders carefully. (Though he may have written down his observations when he was somewhat older, at least some of this observation likely took place during his childhood in rural Connecticut.) He first records observations made of their webs in late afternoons by positioning himself so that the edge of a building just hides the disk of the sun from his eye while allowing him to see the "multitudes of little shining webs and glistening strings of a great length, and at such a height as that one would think they were tacked to the sky by one end" (154–55), strings that carried aloft individual spiders at their ends. The material of the web, he reasoned, must be lighter than air and must be somehow spun from the spider's body. To observe this more closely, he gathered spiders resting on twigs and shook them as he watched them. The spiders did indeed emit a strand from their tails, as he puts it. The language with which Edwards describes what happens next is purely the language of science:

---

[4] See *Works of Jonathan Edwards 6: Scientific and Philosophical Writings*, ed. Wallace E. Anderson (New Haven: Yale University Press, 1980), 147–50. All discussion of Edwards as a scientist and philosopher must now refer not only to the authoritative texts presented in this volume, but to the editor's definitive introductory materials. In this essay I quote from "Of Insects," the first of three "Spider" papers attributed to Edwards. It was used by him as the basis for a letter addressed to a Fellow of the Royal Society of London. This letter exists both in a draft and in its more recently discovered final form. All subsequent references in this essay to scientific or philosophical works by Edwards will refer to texts contained in this volume, indicated as *Works 6*.

if the spider has hold on so much of a web that the greater levity of all of it shall more than counterpoise the greater gravity of the spider, so that the ascending force of the web shall be more than the descending force of the spider, the web, by its ascending, will necessarily carry the spider up unto such a height, as that the air shall be so much thinner and lighter, as that the lightness of the web with the spider shall no longer prevail.   (156–57)

The young Edwards used similar experimental means to investigate the physical properties of rainbows. In response to the proposition that rainbows are the product of sunlight reflecting off clouds, instead of (as Edwards thought) shining through water droplets, Edwards undertook an "ocular demonstration" by "taking a little water into my mouth, and standing between the sun and something that looks a little darkish and spirting of it into the air so as to disperse all into fine drops; and there will appear as complete and plain a rainbow, with all the colors, as ever was seen in the heavens" (298).[5] He also addresses the problem of the rainbow's characteristic circular shape. "To resolve this," he declares, "we must consider this one law of reflection and refraction, to wit: If the reflecting body be perfectly reflexive, the angle of reflexion will be the same as the angle of incidence; but if the body be not perfectly so, the angle will be less than the angle of incidence" (299). This declaration is accompanied by a rather complex diagram of line segments and angles, whose "bare consideration," Edwards asserts, "will be enough to convince any man" (299).

Edwards begins his essay "Of The Rainbow" by declaring that the account he is about to give will be "satisfactory to anybody, if they are fully satisfied of Sir Isaac Newton's different relexibility and refrangibility of the rays of light" (298). Edwards's reading of Newton's *Optics*, as well as other contemporary scientific works, similarly informs a series of "Things to be Considered and Written

[5] "Of the Rainbow," *Works 6*.

fully about" that Edwards compiled while a student.[6] One entry, for instance, seeks to answer why "thunder that is a great way off will sound very grum, which near is very sharp" (222) by considering the motion of sound waves over increasing distance. In another entry, a consideration of the motion of light waves as they arrive from the stars, as distinct from the planets, leads Edwards to the matter-of-fact conclusion that "the fixed stars are so many suns" (237). This conclusion produces two corollaries: 1) "that our sun is a fixed star is as certain as that any one particular star in the heavens is one"; and 2) "'Tis as probable that the fixed stars have systems of planets, as it would be that ours had" (237).[7]

All of this is perfectly in keeping with the spirit of the age of Benjamin Franklin, not to mention the age of Galileo. Yet we find Edwards preaching sermons that hardly seem the product of the same mind. A sermon on "The End of the Wicked Contemplated By The Righteous," for instance, makes the startling assertion that the suffering of the wicked will be a source of pleasure for their loved ones in the next world. Far from your torments in hell being an occasion of grief to the saints in heaven, Edwards explained to the unregenerate in his congregation, those in heaven, whether parents or husbands or wives or brothers or sisters or children or friends, "when they shall behold you with a frightened, amazed countenance...when they shall see you turned away and beginning to enter into the great furnace, and shall see how you shrink at it, and hear how you shriek and cry out; yet they will not be at all grieved for you, but at the same time you will hear from them

---

[6] *Works* 6.
[7] Clarence H. Faust, "Jonathan Edwards as a Scientist," *American Literature*, I (1930, 393–404), points to other entries in Edwards's journals that do not support a view of Edwards as possessing a modern scientific mind. Faust is answered by Theodore Hornberger, "The Effect of the New Science upon the Thought of Jonathan Edwards," *American Literature*, IX (November, 1937, 196–207), who argues that, while many of Edwards's premises derive from Scripture and the medieval tradition, his attitude is nevertheless scientific, declaring that "there is abundant evidence that Edwards possessed in a degree uncommon to his age the scientific mind, which seeks always to relate particular events to general laws." The two articles, together with other essential texts, both primary and secondary, can be found in *Jonathan Edwards and the Enlightenment*, ed. John Opie (Lexington, Massachusetts: D.C. Heath, 1969).

renewed praises and hallelujahs for the true and righteous judgments of God, in so dealing with you."[8] It's difficult to imagine
upon what scientific observation Edwards might have arrived at
this conclusion.

Yet observation is at most only half of the scientific process.
With a mind inclined to Locke as well as to Newton, Edwards in
all his writing shows a strong interest in what *must be*, by necessity.[9] We apply reason to what we observe and arrive at understanding. Thus science can speak about the past and the future,
as well as the present moment. In a striking passage in his "Notes
on Mind,"[10] also written during his student days, Edwards writes

> If a ball of lead were supposed to be let fall from the
> clouds and no eye saw it till it got within ten rods of the
> ground, and then its motion and celerity were discerned
> in its exact proportion, if it were not for the imperfection
> and slowness of our minds, the perfect idea of the rest of
> its motion would immediately and of itself arise in the
> mind, as well as that which is there. So, were our
> thoughts comprehensive and perfect enough, our view of
> the present state of the world would excite in us a
> perfect idea of all past changes. (354)

That is, full comprehension of the present carries with it comprehension of what must have been so in the past to bring the world
to this particular arrangement. By extension, a full comprehension
of the present, as with the present moment of the falling ball of
lead, will also comprehend what must be its future motion.

---

[8] *The Works of President Edwards*, IV, ed. E. Williams and E. Parsons (London: 1817), 517.
[9] The youthful enthusiasm for Locke's *Essay Concerning Human Understanding* which
Edwards recounted shortly before his death is recalled by Samuel Hopkins in his *The Life and
Character of the Late Reverend Jonathan Edwards: A Profile* (Boston, 1765). The strongest
statement of Locke's influence on Edwards remains Perry Miller's *Jonathan Edwards* (New
York: William Morrow and Co., 1949). A strong demurrer is provided by Norman Fiering,
Jonathan Edwards's Moral Thought and Its British Context (Chapel Hill: University of North
Carolina Press, 1981). A more balanced assessment is provided by Anderson.
[10] *Works 6*.

Edwards's scientific inclinations led him to embrace atomism as well, though he resisted the mechanistic materialism of a Thomas Hobbes.[11] Edwards's acceptance of the notion that the world consists of atoms and his assertion that the motion of these atoms defines the nature of the universe strikes us even today as remarkably modern. The motion of even "the least atom," Edwards speculates in "Things to be Considered and Written fully about",[12] "has an influence on the motion, rest, and direction of every body in the universe" (231). The familiar "butterfly effect" of modern chaos theory is simply a restatement of this assertion. But for Edwards this was not a manifestation of a blindly mechanistic universe, but rather a description of God's original act of creation, an act of creation quite different in detail from the account of the creation of the world in Genesis. It is instead startlingly modern in its assertion of a God-originated Big Bang. Edwards continues his meditation on the motion of atoms to a consideration of

> the great wisdom that is necessary in order thus to dispose every atom at first, as that they should go for the best through all eternity; and in the adjusting by an exact computation, and a nice allowance to be made for the miracles which should be needful. And then to shew how God, who does this, must necessarily be omniscient and know every thing that must happen throughout eternity. (231)[13]

A return to "Of Insects," his youthful study of the flying spiders, shows this inclination of Edwards's mind, and its influence on his imagination. When the levity of the spider's web is greater than

---

[11] See Anderson 53–68 for a discussion of Edwards's relation to Hobbes and to mechanistic materialism.

[12] *Works*, 6.

[13] Bruce Kucklick, *Churchmen and Philosophers: From Jonathan Edwards to John Dewey*, (New Haven: Yale University Press, 1985), 12, points out the influence of Newtonian science in the development of a stricter doctrine of determinism among Calvinists in general at the end of the seventeenth century.

the gravity of the spider, Edwards asserted, then the spider must "necessarily" rise in the air, just as a log of lesser density than the surrounding water will necessarily rise to its surface. At the conclusion of his remarks on the spider, he extends this principle of necessity to the ultimate destruction of the spider, and indeed of the majority of all flying spiders, in the vastness of the ocean:

> I say then, that by this means almost all the spiders upon the land must necessarily be swept first and last into the sea. For we have observed that they never fly except in fair weather; and we may now observe that it is never fair weather, neither in this country nor any other, except when the wind blows from the midland parts, and so towards the sea....and they keep flying all that while towards the sea, [they] must needs almost all of them get there before they have done. (160)

Edwards then notes this tendency in all flying insects and concludes that

> without any doubt, almost all manner of aerial insects, and also spiders which live upon them and are made up of them, are at the end of the year swept and wafted into the sea and buried in the ocean, and leave nothing behind them but their eggs for a new stock the next year. (160–61)

A similarly Lockean extrapolation of what must be from what is observed can be seen in Edwards's seemingly peculiar conclusion that those in heaven will rejoice at the suffering of those they once loved. That love exists in this world between those who will go to heaven and those who will not was evident to Edwards, as natural affections exist independently of the love of God and Christ. And in this world this love is often a source of anguish for those beloved of God, as they see their loved ones apparently destined for perdition. Yet those in heaven are free from anguish, for they exist then in perfect happiness. Thus it must necessarily be that the torments of the wicked in hell are no occasion of grief for the saints in heaven, however much our present emotions urge us to feel otherwise. Reason tells us it must be so. Edwards's message to his parishioners was that they must put aside what they would like to think, and rather accept what observation and reason tells us must be so. The truths of religion, no less than the truths of everyday existence, adhere to necessary laws that conform neither to our wishes nor to the limits of our understanding, but to the will of that omnipotent God that has brought the world into being. Far from such a view being at odds with a scientific world view, it is more likely a product of that world view.

If we return, then, to "Sinners in the Hands of an Angry God," we can see that its great emotional power derives not simply from images of torment and destruction, but from the naturalistic reason that propels it towards its conclusions, just as reason, combined with observation, discerns the path of the falling ball of lead and discerns the ultimate destruction of the flying insects. Edwards famously delivered the Enfield sermon not with the exaggerated gesticulations and exhortations of a George White-field, but with a calm and steady matter-of-factness. This tone is set immediately in the choice of text from Deuteronomy: "Their foot shall slide in due time." This is a statement not of exhortation nor judgment, but only of scientific certainty. Given the force of gravity, the slipperiness of footing, and the passage of time, *Their foot shall slide in due time*. It is merely a statement of inevitability, of what must be. A force akin to Newtonian gravity is repeatedly linked to the ultimate end of human souls. "Unconverted men walk over the pit of hell on a rotten covering," Edwards declares, "and there are innumerable places in this covering that are so weak that they will not bear their weight and these places are not seen" (53). It is not something we do, and thus that we might avoid doing, that carries us down to destruction, but only our own human weight, just as the lead ball falls inevitably toward the ground if the force of gravity is not counteracted by some greater upward force.

Your wickedness makes you as it were heavy as lead, and to tend downwards with great weight and pressure towards hell; and if God should let you go, you would immediately sink and swiftly descend and plunge into the bottomless gulf, and your healthy constitution, and your own care and prudence, and best contrivance, and all your righteousness, would have no more influence to uphold you and keep you out of hell, than a spider's web would have to stop a falling rock. (55–56)

The force of gravity cannot itself be seen, though it acts at great distances. It is none the less real for that, nor any less powerful. Science reveals to us forces that cannot themselves be seen except in their effects, which are nonetheless real and irresistible. So too with the great force of ultimate destruction. "The arrows of death fly unseen at noonday," Edwards declares, "the sharpest eye cannot discern them" (53). The sovereignty of this force is like the sovereignty of the prevailing wind, the sovereignty of gravity.

Locke's model of cause and effect also informs Edwards's most famous treatise, his *Careful and Strict Inquiry into the Modern Prevailing Notions of That Freedom of the Will Which is Supposed to be Essential to Moral Agency, Vertue and Vice, Reward and Punishment, Praise and Blame.* The key word here is, of course, "supposed." Edwards by no means concedes that free will is essential to moral agency. Rather the concept of free will is shown by Edwards to be absurd. Each choice we make presupposes a prior choice, a regression which must ultimately lead to a predisposition which logically cannot be of its own choosing. We choose this or that, but when did we choose to be the person who makes the choices we make? Nothing, including moral agency, can bring itself into being. "If once it should be allowed, that things may come to pass without a cause," Edwards argues,

we should not only have no proof of the being of God, but we should be without evidence of the existence of anything whatsoever, but our own immediately present ideas and consciousness. For we have no other way to prove anything else, but by arguing from effects to causes: from ideas immediately in view, we argue other things not immediately in view: from sensations now excited in us, we infer the existence of things without us, as the causes of these sensations: and from the existence of these things, we argue other things, which they depend on, as effects on causes.[14]

"It is indeed repugnant to reason, Edwards concludes, "to suppose that an act of the will should come into existence without a cause, as to suppose the human soul, or an angel, or the globe of the earth, or the whole universe, should come into existence without a cause" (185). In this denial of free will, however, reason might be thought in conflict with our everyday experience of free will, in which case the scientifically inclined might be forced to choose observation over reason. Samuel Johnson, another of Edwards's contemporaries, says as much when he declares that all reason is against freedom of the will, but that all experience is for it.[15] Edwards would not concede this, however, as direct observation as well as reason supports his conclusion. True virtue, true moral agency—and thus the only meaningful acts of will—can be the product of only one thing: love of God and Christ and all things holy. But some experience this love while others do not, though all have the same opportunity. Many will sit in church, hear sermons, read the Bible, and have all about them the evidences of God, and yet not feel love toward God; this is so even of many who intellectually believe. Those who do love cannot be said to have chosen to, while the rest have chosen not to, for what made some choose one way and some choose the contrary? When did

[14] *Works of Jonathan Edwards 1: Freedom of the Will*, ed. Paul Ramsey (New Haven: Yale University Press, 1957), 183.
[15] James Boswell, *Life of Johnson*, ed. R. W. Chapman, 3rd ed., corr. J.D. Fleeman (Oxford: Oxford University Press, 1970), 947.

each individual choose to be the chooser making his or her choices?[16]

In his *Treatise on Religious Affections*, Edwards makes this point with an analogy to readers of Milton. Of the many who read Milton, he notes, only a few will really sense his beauty, and it is impossible either to guess who these few will be or to explain why they are. Without thinking much about election and free will, English professors observe this every day. Some students apprehend the beauty of Milton and others do not. This is not usually determined by who works the hardest or who comes from an educated background (though their success in writing "A" papers might well be influenced by such factors). Nor can we say that those who do do so because of our lectures and discussions, though we lecture and discuss in the hopes of bringing this to pass. Rather, something else determines this, something other than either the will of the student or the efficacy of the teacher.

Yet if there is an effect there is a cause. For Edwards, the final cause, the cause that precedes all, is God. And what, Edwards asks, is the alternative? A contingent world that has come into existence out of nothing? Nothing in science can comprehend such a singularity, though a backward reasoned Lockean train of cause and effect might take us all the way back to the moment of singularity. For Edwards, the first cause is God. But if God is the cause of all, why has he not imparted to all the love that he has imparted to some? However philosophically one might answer this, objectively one must concede that he has not. And thus the great tragic doctrine of Calvinism—that some are chosen by God for glory and others are not—is arrived at not by dogma, but by reason and observation. The fact that we object to such a state of

---

[16] John E. Smith, *Jonathan Edwards: Puritan, Preacher, Philosopher* (Notre Dame: University of Notre Dame Press, 1992), 28, makes this point: "The curious fact is that Edwards, wittingly or not, was actually maintaining a theory of *self*-determination, except that, due to the doctrine of election, the individual has no hand in determining the *nature* of the self that is to do the determining." But it isn't necessary to say that the doctrine of election exists first in Edwards's mind; it may as well be that observation of how things actually are led him to affirm the doctrine of election.

things is of no more consequence than a wish that the laws of gravity might be different than they are.

For Edwards, gravity was a continuous manifestation of God's will, observable in the behavior of objects but not a result of anything inherent in those objects themselves. Similarly, though in the spiritual rather than physical realm, the upward pull of what Edwards called divine light has an effect on souls that might be thought of as the opposite of the downward gravitational pull invoked in "Sinners in the Hands of an Angry God." Among Edwards's most beautiful sermons is that on "A Divine and Supernatural Light." This light, imparted directly to the soul by God, is of a different sort than natural light, but is no less real for that, just as gravity is not the less real for being unseen. We know of the existence of both by observation of their effects. The effect of gravity draws earthly objects inevitably downward. The effect of the divine and supernatural light of God, on the contrary, draws the soul irresistibly toward God. The effect of God's divine light on the individual soul is not primarily rational; nevertheless, its effects can be rationally observed. Without this light imparted to the soul by God, the mind of man is capable of understanding that God is good, but rational understanding alone is not sufficient for true virtue. Rather, a sense of the heart is needed,

> as when the heart is sensible of pleasure and delight in the presence of the idea of it. In the former is exercised merely the speculative faculty, or the understanding, in distinction from the will or disposition of the soul. In the latter, the will, or inclination, or heart, are mainly concerned.[17]

Thus Edwards, though himself endowed with a rational mind equal to if not superior to that of a Ben Franklin, nevertheless gives primacy to emotion rather than reason in human salvation. His

---

[17] "A Divine and Supernatural Light," *Works of Jonathan Edwards 17: Sermons and Discourses 1730–1733*, ed. Mark Valeri (New Haven: Yale University Press, 1999), 413–14.

reason for doing so is entirely rational, however: his rational observation of the importance of love in saving virtue.

The love that Edwards has in mind, however, is not to be confused with self-love, which Edwards in his notes on "The Mind" declares not to be properly called love at all.[18] Yet self-love is the natural condition of human beings. If we are to attain love of God, love of Christ, love to being in general—that is, if we are to achieve that love through which is found saving virtue—our natural self-love must be obliterated. But how is this to be achieved? Again, Edwards finds his answer in observation. Just as true love of God is not so much a matter of rational understanding as it is a matter of a transforming emotion, so too is the negation of self-love a matter of a transforming emotion: the heart must be purged of self-love through terror before the sweetness and delight of true love can be achieved.[19] In Edwards's "Personal Narrative," he recalls "two early seasons of awakening, before I was met with that change, by which I was brought to those new dispositions, and that new sense of things, that I have since had."[20] In describing these two unsuccessful awakenings, he declares that, though he had made "seeking my salvation the main business of my life," yet "it never seemed to be proper to express my concern that I had, by the name of terror" (791). Yet not only does he come to accept that which formerly was terrible to him, he comes to find the greatest delight in these things. "The appearance of everything was altered," he recalls, even the appearance of thunderstorms:

> And scarce anything, among all the works of nature, was
> so sweet to me as thunder and lightning. Formerly,
> nothing had been so terrible to me. I used to be a person
> uncommonly terrified with thunder: and it used to strike

---

[18] *Works*, 6, 337.

[19] The second stanza of the familar gospel hymn "Amazing Grace" makes this point: "Twas grace that taught my soul to fear/And grace my fears relieved."

[20] "Personal Narrative," in *Works of Jonathan Edwards 16: Letters and Personal Writings* ed. George S. Claghorn (New Haven: Yale University Press, 1998), 790.

me with terror, when I saw a thunderstorm rising. But now, on the contrary, it rejoiced me. I felt God at the first appearance of a thunderstorm. And used to take the opportunity at such times, to fix myself to view the clouds, and see the lightnings play, and hear the majestic and awful voice of God's thunder: which often was exceeding entertaining, leading me to sweet contemplations of my great and glorious God. And while I viewed, used to spend the time, as it always seemed natural to me, to sing or chant forth my meditations: to speak my thoughts in soliloquies, and speak with a singing voice. (793–94)

The contrast with Benjamin Franklin is striking: Franklin, the man of science, stands in the thunder and flies a kite; Edwards sits in the thunder and chants. But does this mean, finally, that Edwards, at least at such moments, is not a man of science? After all, his notebooks are full of speculations on the sound waves of thunder—how they move through space, how they are changed by distance, etc. Can they be such mechanical phenomena and at the same time be the majestic and awful voice of God? Or are these responses to thunder mutually exclusive? Do we find in Edwards a voice not unlike a narrator in a story by Poe, a voice that seems rational and indeed declares itself to be rational but is in reality utterly irrational?

Perhaps not. In the same notebooks, Edwards speculates on the origins not just of will or affections, but of the universe itself. Preexistent and eternal, he reasons, must be space; we can imagine a universe empty of all else, but not empty of space itself. (Modern physics would have us imagine space itself coming into being at the moment of singularity, but though we can say these words and explain the concept to an inquisitive child, yet we cannot help but picture even this singularity as occurring in some kind of empty space.) "It is self evident I believe," Edwards writes, "to every man that space is necessary, eternal, infinite, & Omnipresent. But I had as good speak Plain, I have already said as much as that Space is God."[21] That is, speaking plainly, God is not literally a bearded man in a robe; God is the eternal pre-existent, the infinite upon infinite that is both the ground and cause of all, and the eternal mind that comprehends all, without which nothing would exist, just as, in Newton's *Optics*, colors do not exist without a mind to see them. If for a moment, Edwards reasons, all minds would cease to exist, the mind of God as well as the minds of man, at that moment the universe itself would cease to exist. To think on this is not to be unscientific, but rather to think science's thoughts through to the profound mystery of existence. God is being, existence itself, that which must be; the alternative to God, to existence, is nothingness, inconceivable negation. What mind, what heart, what imagination, conceiving of these alternatives in their fullness, would not be filled past bearing first with terror, at the imagination of utter contingency and unspeakable nothingness, and then with joy, at the comprehension of existence itself in all its unutterable sweetness?

The sermon at Enfield, then, might be thought of as a kind of experiment—an experiment in terror and its effect in awakening the heart to the awful sovereignty of God. Edwards himself called true religion *experimental religion*. By this he meant that religious knowledge must be experienced. But not every mind is going to be capable of experiencing the terror and joy of existence reading Newton's *Optics* and Locke's *Essay Concerning Human Understanding* and contemplating the implications these works might suggest, though Edwards may well have had such a mind. Rather, poetic images of the awful sovereignty of that which *is* might be more efficacious: images of a spider dangling by a thread over the fire, of the arrows of death flying unseen at noonday, of the rotten covering of nothingness through which we will fall if we do not come to love the awful mystery of existence, if we instead give ourselves to over to mere contingency and nothingness.

Edwards's imagination, unlike Franklin's, was not less poetic for its profound inclination towards a scientific way of thinking.

---

[21] "Of Being," *Works* 6, 203.

Rather a new tragic vision emerges, beyond the capacity of a Franklin, in which the mere weight of our human nature carries us down to destruction, in which things as they are *are* as they are, in which cause and effect produce necessity, in which if all could be seen clearly, all that has been and all that must be would be seen. Edwards's vision is in part a tragic vision of a falling ball of lead. To think of nothing, Edwards writes, we must think the same as the sleeping rocks dream of.[22] Yet we also can apprehend the great wonder of existence, like the taste of honey in our mouths, and love that existence, that being, with all *our* being.

Religious fundamentalism is much on our minds these days, and it has become a commonplace to suggest that the advent of religious fundamentalism is a reaction against the advent of science. And this is probably so. But the example of Edwards suggests that the opposite might also, to some extent, be true. Much has been written on Edwards's acceptance of the doctrine of election in his college days. From his childhood up, Edwards writes in his "Personal Narrative," his mind had been "full of objections against the doctrine of God's sovereignty, in choosing whom he would to eternal life, and rejecting whom he pleased" (791–92). But then, Edwards continues, his mind changed, and he became convinced that this doctrine was true and just, though he could not explain afterwards how this change had taken place. Commentators who assume Calvinism to be antithetical to modern scientific thought suggest that this conversion brought with it the end of Edwards's scientific thinking. But it is perhaps just as convincing to argue that this conversion was the result of Edwards's scientific thought, rather than its negation, particularly as scholarship places the writing of his scientific papers closer to the end of his college career. In Edwards we find a religious mind that has taken to heart certain scientific principles—that the world operates in accordance with laws that are everywhere the same, for instance, or that we must recognize the truths that are rather than the truths we'd like. We can not each of us simply

---

[22] "Of Being," *Works* 6, 203.

make up our own laws of thermodynamics. If God is one thing, he cannot be something else, anymore than science in Russia can be something different from science in Chicago.[23] Science tells us things are uniform, that nature is nature, that observation and reason can lead us toward certainty. In this sense, liberal theology and ecumenism might better be understood as the reaction against the great burden of the scientific mind, a way of thinking in which there are no irrevocable laws, and certainly there is nothing predetermined. In liberal theology, we can each have things our own way; fundamentalists, on the other hand, think no, we cannot have things as we would like them to be, but must understand and accept things as they are. Many such thinkers since Edwards are no doubt less imaginative than Edwards and they may on the surface (and even at their conscious core) be more hostile to such scientific concepts as evolution, but they too assert a God that is one way and not another, a God who must be consistent just as the laws of nature are consistent and predictable. They believe in the great power of God, they may even believe in miracles, but they don't believe either in magic or in a God defined by the needs and desires of individual believers.

---

[23] Recent postmodern attempts to question the universality of scientific principles can be seen as a more recent liberal manifestation of the burden scientific thought places on spiritual consciousness.

"THE FURNACE OF AFFLICTION"
Anne Conway, Henry More, and the Problem of Pain

Anna Battigelli
State University of New York at Plattsburgh

The merest schoolgirl, when she falls in love, has Shakespeare or Keats to speak her mind for her, but let a sufferer try to describe a pain in his head to a doctor and language at once runs dry.
— Virginia Woolf, "On Being Ill"[1]

Among the many physicians to whom Anne Conway submitted herself in an effort to find a cure for the excruciating and debilitating pain that plagued her throughout her life, the only one to offer some relief was Francis Mercury van Helmont. The physical relief he brought her was only partial and temporary, yet she so valued his company that he came to live at Ragley Hall in Warwickshire for most of the last nine years of her life. Her association with van Helmont came at the cost of departing philosophically and theologically from the most devoted

---

[1] Virginia Woolf, *The Moment and Other Essays* (New York: Harcourt, Brace 1948), 11.

of friends, the Cambridge Platonist Henry More. Under van Helmont's influence, Conway abandoned orthodox Christianity to become heterodox in two related ways: like van Helmont, she appropriated Kabbalistic thought into her natural philosophy, thereby rejecting More's dualistic theory of matter; and, like van Helmont, she became a Quaker, thereby rejecting Henry More's rational theology. At the heart of Conway's rejection of More's natural philosophy and theology was the problem of pain. The narrative of her break with More highlights David Morris's observation that "pain, whatever else philosophy or biomedical science can tell us about it, is almost always the occasion for an encounter with meaning."[2]

Conway suffered incapacitating headaches from the age of twelve onward. Her "fitts," as she called them, were described by Thomas Willis, one of the several eminent physicians to treat her unsuccessfully:

> This sickness being limited to no one place of the Head, troubled her sometimes on one side, sometimes on the other, and often thorow the whole compass of the Head. During the fit (which rarely ended under a day and a nights space, and often held for two, three, or four days) she was impatient of light, speaking, noise, or of any motion, sitting upright in her Bed, the Chamber made dark, she would talk to no body, nor take any sleep, or sustenance. At length about the declination of the fit, she was want to lye down with an heavy and disturbed sleep, from which awaking, she found her self better, and so by degrees grew well, and continued indifferently well.[3]

---

[2] David B. Morris, *The Culture of Pain* (Berkeley: University of California Press, 1991), 34.
[3] Thomas Willis, *Two Discourses Concerning the Soul of Brutes, Which is that of the Vital and Sensitive Man* (London, 1683), 121–22.

At first these episodes recurred every month or twenty days, but their frequency increased until soon she was seldom free from pain.[4]

As others have noted, Conway exhausted every option that early modern science offered in her search for a cure.[5] She had access to the best physicians of her day: Theodore de Mayerne, Thomas Willis, William Harvey, and Robert Boyle. She experimented with healers, such as Matthew Coker and Valentine Greatrakes. Her friends and family members scoured Europe for possible aid, and their correspondence is punctuated by hopeful suggestions for powders, purges, diets, and rest cures. She tried baths, spa waters, mercury, and blood-letting.[6] As Willis recounts, "there was no kind of Medicines...which she took not, from the Learned and the unlearned, from Quaks, and old Women."[7] None of these cures worked, and each onset of illness left her increasingly weakened. In 1656, she traveled to France with Henry More in order to undergo the desperate measure of trepanning–an operation that would have opened her cranium–but the doctors decided against it, much to More's relief. They did, however, open her jugular veins, a dangerous procedure, which once again failed to bring her respite from her pain. In 1660, at a time when much of the rest of the nation was occupied with the Restoration of the king, Conway suffered an additional blow in the loss to smallpox of her only son, Heneage Edward Conway, who was not yet three years old.[8] This loss contributed a spiritual dimension to the

---

[4] Willis, *Two Discourses*, 122.

[5] The literature on Conway's illness begins with Willis, *Two Discourses*. A more recent clinical analysis is provided by Gilbert Roy Owen, "The Famous Case of Lady Anne Conway," *Annals of Medical History* 9:6 (1937): 567–71. Sarah Hutton provides an overview of Conway and early modern medicine in "Of Physic and Philosophy: Anne Conway, F.M. van Helmont and Seventeenth-Century Medicine," In *Religio Medici: Medicine and Religion in Seventeenth-Century England,* ed. Ole Peter Grell and Andrew Cunningham (Aldershot: Scolar Press, 1996), 228–46. More recently, Peter Loptson disagrees with Owen's analysis: see *Anne Conway, The Principles of the Most Ancient and Modern Philosophy*, ed. Peter Lopson (Delmar, New York: Scholars Facsimiles & Reprints, 1998), 13–14.

[6] Willis, *Two Discourses*, 122.

[7] Willis, *Two Discourses*, 122.

[8] Heneage Conway was born on February 6, 1658, and died on October 14, 1660. See Nicolson, 123, 126.

physical pain with which she was by now all too familiar. "It hath pleased God," she wrote to More,

> to exercise me by divers afflictions and by one so sensible in the death of my child, that you must not wonder if I tell you it hath extorted from me a griefe proportionable to so great a losse. Neither am I yet able so well to overcome it as might beseeme one whose present constitution...ought not to be so much affected with such injoyments of this world.[9]

Conway herself became ill and was expected to die; her recovery, she lamented to More, denied her that "perfect release to all my sufferings" (*Conway Letters*, 181). By 1664, after a particularly long illness, her struggle with pain and sorrow seems to have exhausted her, and she freely admitted to being weary of life: "I cannot dissemble so much as not to professe myself very weary of this condition. I pray God enable me with patience to bear whatsoever my sad fate hath designed for me and give me that entire resignation to his will which I endeavour after and do stand in so much need of," she wrote again to More (*Conway Letters*, 224).[10] Her health had all her life precluded entry into the social world to which she was by her wealth and status admitted; during most of the last nine years of her life, she was confined to her bedroom. Studious by nature, she claimed to prefer solitude, and she spent most of her life alone with her studies and with her pain.

---

[9] Marjorie Nicolson, *The Conway Letters: The Correspondence of Anne, Viscountess Conway, Henry More, and their Friends, 1642–1684*, ed. Sarah Hutton, rev. ed. (Cambridge: Cambridge University Press, 1992), 181. Unless otherwise noted, all subsequent citations to letters between members of the Conway/More circle are to this edition and will be provided in the text.

[10] For Sarah Skwire, this 1664 letter indicates a turning point in Conway's conception of pain in that she began to perceive her illness less as an interruption than as a defining characteristic of her life; she understood that she was "a sick person, a sufferer." Her distress at recovering her health in 1660, however, may suggest that she experienced physical and emotional pain as a defining characteristic earlier, perhaps triggered by the death of her son.

See Sarah E. Skwire, "Women, Writers, Sufferers: Anne Conway and Ann Collins." *Literature and Medicine* 18.1 (1999), 1–23. See particularly page 10.

She was encouraged in her confrontation with suffering by the devoted and lifelong correspondence of More, who tutored her and corresponded with her from 1650 onward. Knowing that physicians had failed and were unlikely to bring Conway relief, More stuffed his letters with learned and rhetorically elegant variations on the theme of Christian resignation. In 1654 he wrote:

> The best advice that I can give you, is that which I endeavour after myself as much as I can, that your phansy add nothing to the torture of your sense, and to resigne yourself wholy to the will of God, useing the best meanes you can for the removall of your present affliction....death itself which seemes the most terrible thing in the world is but a returne to greater freedom. (Conway Letters, 100).

Two years later, he added that "All thinges whatsoever tend to the good of those that are good even those that seeme for the present to be the greatest evills. For the pious and vertuous come out of them, as gold purify'd out of fire" (Conway Letters, 154). Accordingly, he urged her, "Submit yourself to God and...endeavour to say heartily those words of Job: It is the Lord: lett him do what it pleaseth him" (Conway Letters, 131). In 1658 he attempted to console her by remarking that "the affaires of the world are but the working of the clouds, if we cast but our eyes of[f] them awhyle and then look on them againe, they are not the same" (Conway Letters, 386). In this transitory life, pain itself, he advised, offers an opportunity for humility and for the resignation of one's will to God:

> Certainly we are in a capacity by affliction, of impressing such characters on our spiritt, of perfect humility and resignation of ourselves unto God, as neither before or when it is over, without a miracle we can be. And therefore it is good striking whyle the iron is hott, and

> inuring ourselves to fayth in God and dependence on and
> full submission to him, whyle we are thus moulded in
> the furnace of affliction. (*Conway Letters*, 154)

Pain, he argues again and again, is helpful in acquiring that perfect
resignation to divine will that is so difficult to attain.

The problem of pain provides a key theme in the correspondence between More and Conway. More's letters to Conway not
only argue for the spiritual utility of pain; they also reveal the
degree to which his friendship with Conway caused him emotional pain. For instance, after a visit to Conway at Kensington in
1656, More returned to Cambridge, where by all accounts he had
lived happily since becoming a Fellow fifteen years earlier in 1641,
but instead of finding happiness in his familiar rooms, he found
that he had been happier still in Conway's company.[11] In a long
and revealing letter to Conway, he complained of being

> depriv'd of the greatest enjoyment this World ever
> afforded me which is the pleasure of your Ladiships
> company from which I reape so great satisfaction, that it
> makes that Life, which used to give me the greatest
> content, very dead and heartless to me. When I first
> came into my chamber, methought it look'd very sad and
> desolate, and I found it no complement, that after my
> converse with so noble a friend, my retirement to Cam-
> bridge would be like coming into an obscure cottage.
> After a long walk in the sunshine and snow all things
> afterwards looke darke. I profess, Madame, I never knew
> what belonged to the sweetness of friendship before I
> mett with so eminent an example of that virtue....I can
> not command my self from most affectionately loving
> her, whom it is my duty...to honor and adore. (*Conway
> Letters*, 128–29)

[11] For information on More's early years at Cambridge, see C. C. Brown, "Henry More's
'Deep Retirement': New Material on the Early Years of the Cambridge Platonist," *The Review
of English Studies*, ns. 20 (November 1969): 445–54.

He turned to her, he told her in another letter, "as naturally as the needle turnes to the North" (*Conway Letters* 162). He claimed to be "nothing but an Aggregate of my friends so that they that are the best and choicest of them are the greatest part of myself: and that you are the chiefest of all I must with thankfulnesse acknowledge, nor can anyone deny it" (*Conway Letters*, 165). Elsewhere, he noted, "I find no where more serious, more sutable conversation than with your Ladiship" (*Conway Letters,* 355).

His attachment to her could also be discomfiting: when, after the Restoration, he learned that the Conways were to leave for Ireland for two years, during which he would be deprived of his visits to her, More became so distraught that he left her in haste without properly taking leave, only to apologize immediately upon his arrival in Cambridge for leaving "so abruptly by halfes." Had he traveled on horseback rather than in a coach which obliged him to other travelers, he added, he "should have found out some occasion after I had been half way of my first dayes journey to returne and take leave of your Ladiship once more" (*Conway Letters,* 183). Though he did not experience physical pain to the degree that Conway did, his elegant arguments urging her to make use of "the furnace of affliction" suggest his own familiarity with emotional and spiritual pain.

However evident it is that More loved Conway, it is important to see that theirs was an intellectual collaboration as well as a friendship. More's claims that Conway was "the chiefest" part of his sense of self and that he found in no one else "more serious, more sutable conversation" suggest this. Evidence of their collaboration is also present in their extensive correspondence, much of which has been preserved. A. Rupert Hall has suggested that More's disillusionment with Descartes, which can be "dated to the middle 1650s," may reflect Conway's influence. As Hall puts it, "the preoccupations of her mind [may have] pushed his further down the road it might have been inclined to take in any case."[12]

[12] Rupert A. Hall, *Henry More: Magic, Religion and Experiment* (Oxford: Basil Blackwell, 1990): 122.

Richard Popkin goes further to note that "there is a serious possibility that she was really the co-author of *Conjectura Cabbalistica*, the *Antidote against Atheism* and *The Immortality of the Soul*."[13] Their attachment to one another came, in large part, from their shared sense of purpose. The advances of natural philosophy had sought to identify the causes of things within nature as opposed to outside of nature.[14] The effort to banish supernatural causes led increasingly to a vision of the world as wholly mechanistic and material. The extreme proponent of this mechanistic material world view was Thomas Hobbes, who claimed that "the whole masse of all things that are is Corporeall, that is to say, Body."[15] Against the encroachment of materialists like Hobbes, both Conway and More labored to preserve the spirit world. Preserving the spirit world was for them tantamount to preserving Christianity. More claimed in *The Immortality of the Soul* that materialism led to the idea that it was

> impossible there should be any God, or Soul, or Angel, Good or Bad; or any Immortality or Life to come. That there is no Religion, no Piety nor Impiety, no Virtue nor Vice, Justice nor Injustice, but what it pleases him that has the longest sword to call so. That there is no Freedom of the Will, nor consequently any Rational remorse of Conscience in any Being whatsoever, but that all that is, is nothing but Matter and corporeal Motion.[16]

More's *An Antidote against Atheism* (1653), *The Immortality of the Soul* (1659), the *Divine Dialogues* (1668), and *Enchiridion Metaphysicum* (1671) sought to prove the existence of spirit. For her part, Conway went further than More. In *Principles of the Most Ancient*

---

[13] Richard H. Popkin, "The Spiritualistic Cosmologies of Henry More and Anne Conway," in *Henry More (1614–1687): Tercentenary Studies*. ed. Sarah Hutton (Dordrecht: Kluwer, 1990), 100.

[14] See Hall, *Henry More*, 6.

[15] Thomas Hobbes, *Leviathan*, ed. C. B. Macpherson (New York: Penguin, 1985), 689.

[16] Henry More, *The Immortality of the Soul* (London, 1659), 56.

and Modern Philosophy, published posthumously in 1690, she claims that everything is spirit.

If Conway started out as More's "Heroine pupil" and even became his collaborator, her independence of mind led her eventually to depart from More's teaching, causing palpable strain in a relationship that both valued highly.[17] As More once remarked, Conway was "one that would not give up her Judgment entirely unto any."[18] Their differences emerged in the last decade of her life. More posited a dualistic system in which the universe consists of matter and spirit; Conway posited a monistic system in which everything is spirit. He positioned himself cautiously between the twin threats of sectarian enthusiasm and Roman Catholicism; she did away with all such caution and eventually came to embrace the Quakers, a despised sect feared in the seventeenth century for their fanaticism and for their disregard for class order.[19] More lived his entire professional life in Cambridge, where he dove unhesitatingly into print controversy. Apolitical by nature, his language could nevertheless quickly become combative as he countered opponents' arguments. Conway, by contrast, was confined most of her adult life to her house and later to her bedroom due to illness. If More defined himself through an expansively discursive nature, Conway seems to have yearned for something beyond discourse. He embraced the printed page, confronting pain by writing about it as a way of submitting to divine will; she turned to an experiential mode diametrically opposed to his discursive mode, confronting pain with the Quakers in absolute silence.

---

[17] Richard Ward, *Life of Henry More* (1710), ed. Sarah Hutton et al. (Dordrecht: Kluwer, 2000), 117. All citations are to this edition.
[18] Ward, *Life of Henry More*, 121.
[19] See James Walvin, *The Quakers, Money and Morals* (London: John Murray, 1997), 13–27; Richard T. Vann, *The Social Development of English Quakerism, 1655–1755* (Cambridge: Harvard University Press, 1969), 28–32; Bonnelyn Young Kunze, *Margaret Fell and the Rise of Quakerism* (Stanford: Stanford University Press, 1994), 131–42; See also Rosemary Anne Moore, *The Light in their Consciences: Early Quakers in Britain, 1646–1660* (University Park: Pennsylvania State University Press, 2000).

More's discursive nature is evident in the fact that he conceived of his life in terms of the texts that he produced. As his first biographer, Richard Ward, noted in 1711, More claimed to have been "so busie in his Chamber, with his Pen and Lines, as not to mind much the Bustles and Affairs of the World that were Without."[20] It was perhaps in part his bookish disposition that allowed him to remain in Cambridge throughout the tumults of the English civil war and its aftermath "virtually unaffected" by the Puritan purge of the university in 1644, by the Protectorate, the Commonwealth, and Restoration.[21] In the dedication to his 1647 *Poems*, More acknowledged his "bookish disease" (sig. A3r). Ward adds that More's voice, which was thin and high and thus not particularly "fit for...Public Orator[y]" forced More himself to admit that he "should not have known what to have done in the World, if he could not have preach'd at his Fingers Ends" (*Life*, 43).[22]

Writing was without question the defining act of More's life. But his attitude toward writing was complicated by a spiritual crisis of three or four years that began immediately upon his earning his B.A. at Cambridge.[23] After an extended struggle, he

---

[20] Ward, *Life of Henry More*, 117.

[21] Aharon Lichtenstein, *Henry More: The Rational Theology of a Cambridge Platonist* (Cambridge, Harvard University Press, 1962), 8. See also Rupert A. Hall, *Henry More*, 88; C. C. Brown, "Henry More's 'Deep Retirement': New Material on the Early Years of the Cambridge Platonist" 445–54.

[22] For his more recent biographer, Robert Crocker, More's account of his life is significant because his narrative

> becomes an explanatory record of his numerous publications. Rarely does he allow the events of the outside world, even when they directly concern him, to intrude upon his story. It is the story of the publication in successive works–often in direct opposition to "Atheism," "Enthusiasm," and "Superstition" of his age.

See Robert Crocker, "Henry More: A Biographical Essay," in *Henry More (1614–1687): Tercentenary Studies*, 3.

[23] As Aharon Lichtenstein notes "More found that four years of reading Aristotle, Cardan, Julius Scaliger, and other Philosophers of the greatest Note had left him unsatisfied. They failed to deal adequately with the eternal issues which he now encountered face to face–the problems of the nature and destiny of the human soul and of its relation to God" (*Henry*

seems to have had an illumination based on his reading of the *Theologica Germanica*, a fourteenth-century treatise first printed in 1518 under the direction of Martin Luther. For More, its urging that "we should throughly put off, and extinguish our own proper Will; that being thus Dead to our selves we may live alone unto God, and do all things whatsoever by his Instinct, or plenary Permission; was so Connatural, as it were, and agreeable to my most intimate Reason and Conscience, that I could not of anything whatsoever be more clearly or certainly convinced."[24] As a result of this illumination, More claimed to subordinate his "insatiable Desire" for knowledge to the desire for "a more full Union with this Divine and Coelestial Principle, the inward flowing Well-spring of Life eternal" (*Life*, 19–20).

One senses in statements such as this one the affinities between More's theological position and that of Quakers like George Keith, who claimed to have been converted to Quakerism after reading More's *Mystery of Godlinesse*.[25] But Keith and other Quakers professed a confidence in the direction provided by the

More, 5).

[24] Ward, *Life of Henry More*, 19.

[25] More's affinity with the Quakers, whom he both despised and resembled has been noted both by Nicolson and Allison P. Coudert. More's surprise when he learned that George Keith became a Quaker after reading More's *Mystery of Godlinesse* can be found in *Conway Letters*, 341. As Marjorie Nicolson and others have noted, both the Cambridge Platonists and the Quakers shared similarities:

> Surrounded by formalism, both longed for simplicity; in the midst of arbitrary authority, both sought truth in spiritual inwardness. 'The light within'–it is as familiar a phrase in More as in the Quakers. To both of them, also, true religion in those days had been smothered by a mass of irrelevant detail; both sought a return to what they loved to call the "primitive times of the church," to the essentials of Christianity.

See Marjorie Nicolson, "George Keith and the Cambridge Platonists," *Philosophical Review* 39: 36–55. Coudert points out that "certain aspects of More's neoplatonic orientation make his initial attraction to both [Quakers and Kabbalistic doctrine] entirely comprehensible." She notes in particular, his tolerance, his interest in the belief in the perfectibility of man and universal salvation. See Coudert, "Henry More, the Kabbalah, and the Quakers," in *Philosophy, Science, and Religion in England, 1640–1700*. ed. Richard Kroll, Richard Ashcraft, Perez Zagorin (Cambridge: Cambridge University Press, 1992), 34.

inner light that More could not, finally, share. In James Walvin's words, the Quakers

> relegated the importance of the Scriptures in favour of the pre-eminence of this inner spirit, and so rejected the necessity for an educated clergy to lead and interpret. Even the Bible was demoted, to become..."a book like any other." What mattered was not so much biblical stories about Christ and the past, but one's own feelings of the present. Heaven was *within* the Quaker believer.[26]

The Quaker belief that heaven was *within* is evident in William Penn's claim in a letter to Conway that "we preach not our selves, but the light of Christ in the conscience, which is gods faithfull and true witness, that the worldly, pompous church has slain, and made merry over" (*Conway Letters*, 402). Though More acknowledged that "there is something before and better than Reason, whence Reason it self has its rise," he also insisted on the rational nature of Christianity.[27] His concept of "a more full Union" with God thus countered Penn's reliance on the inner light by balancing "feeling" with "learning" and "divine grace" with "discourse."

More's wrestling with reason—defending it as central to Christianity and yet acknowledging the existence of something beyond it—defines his career and informs the qualified mystical rationalism that he defended in volume after volume.[28] It reflects a skeptical habit of mind that avoided dogmatic claims of certainty regarding religious matters. We see an instance of his dual attitude toward reason at work when he complains about those who

---

[26] James Walvin, *The Quakers, Money & Morals*, 13. Walvin quotes Christopher Hill, *The World Turned Upside Down* (London: Penguin, 1975), 214.

[27] Henry More, *A Collection of Several Philosophical Writings* (London, 1662), vii–viii. More is quoting Aristotle.

[28] Samuel Mintz usefully distinguishes More's two uses of reason: on the one hand, More, like the other Cambridge Platonists, seems to understand reason as "clear philosophical thinking;" on the other hand, he also uses "reason" to denote not a faculty capable of *perceiving* but a faculty of *receiving* the supernatural. See Samuel Mintz, *The Hunting of Leviathan* (Cambridge: Cambridge University Press, 1962), 82.

propose some sort of demonstrable proof for the immortality of the soul. Such proof, he argues,

> [would] too forcibly driv[e] men to obedience if they had their immortality as demonstrable as; That the three angles in a triangle are equall to two right angles. Besides it would prevent that fitting trial of the soul, how she would be affected if there were nothing to come.[29]

For More, then, skepticism guarded against the dangers of dogmatism and certainty and helped one engage in the "fitting trial of the soul." Doubt and its discomforts were more efficacious for spiritual development than "Demonstration and Infallibility." In holding to this belief, More was typical of the Cambridge Platonists, who as C.A. Patrides has noted, used the metaphor of the candle evident in Proverbs 20.27–"the spirit of man is the candle of the Lord"–as a point of departure for considering the relationship between reason and revelation.[30] The light of reason could provide useful guidance, but its light could not, finally, banish the darkness of doubt altogether. More's skeptical habit of mind caused him to be wary of relying too confidently on either the inner light or on human reason. Attracted as he may have been to "spiritual inwardness" and to the Quaker determination to restore the primitive church, he stopped short of the Quaker certainty that "the light of Christ" is "in the conscience."[31] He wrote, one might conclude, as much to test and check his own use of reason as to combat the incorrect use of reason by others. When in his letters to Conway he addresses the problem of pain, his focus is generally on the rightness of submitting to divine will rather than on his confidence that pain purges sin.

Conway was a diligent pupil, and she not only resigned herself to a life of pain; she also became a famous embodiment of

---

[29] Henry More, *Philosophical Poems* (Cambridge, 1647), Sig. B3v–B4r.
[30] *The Cambridge Platonists*. Ed. C.A. Patrides (Cambridge: Harvard University Press, 1970), 18.
[31] *Conway Letters*, 402.

Christian resignation and suffering. More had publicly referred to her as "vertue become visible" when he dedicated his *Antidote* to her in 1653.[32] "It seems to me as reall a point of Religious worship to honour the Vertuous as to relieve the Necessitous," he announced in that same dedication, referring again to Conway (*Antidote*, 2). He was hardly alone in praising Conway. The Anglican cleric George Rust praised her as a "glorious" example of "resignation and patience" (*Conway Letters*, 302). Jeremy Taylor wrote his *Christian Consolations* for her. Her step-brother, Heneage Finch, described her as "the living Example of a most Resigned Soule" *(Conway Letters*, 354). A Quaker friend used alchemical imagery similar to More's by describing Conway as a Christ figure, writing "those that come out of this furnace [are] like...Gold mor refin'd...[and are] like an Elixor...to the world" (*Conway Letters*, 280). In 1671, More told her, "your virtues and sufferings hav[e] made you as famous as any one [in] the nation" (*Conway Letters*, 340). Even for those like her brother-in-law Sir George Rawdon, who could not understand her interest in the Quakers, she remained a "gallant soldier" (*Conway Letters*, 433). She had become a figure whose grace, patience, and Christian resignation in the face of unrelieved suffering served as a fit subject for meditation. Had she been merely interested in an explanatory discourse for her daily experience of pain, Henry More's rational theology might have satisfied her.

It proved, however, insufficient. In a final effort to bring Conway some physical relief from pain, More introduced her to Francis Mercury van Helmont in 1670. In a letter to Rawdon, Conway described van Helmont's aid in the following way:

> To your kinde enquiry after what ease I find from Monsieur van Hellmont, I must give this account, my paines and weakness does certainely increase daily, but yett I doubt not, but I have had some releef (God bee thanked) from his medicines, I am sure more then I ever

---

[32] More, *An Antidote Against Atheism* (London 1653), 3.

had from ye endeavours of any person whatsoever else, but yett I have had much more satisfaction in his company, as [he has] yett the patience to continue with mee in my solitude, which makes it the easier for mee, none of my own relations having the leasure to afford mee that comfort, and indeed I think very few friends could have patience to doe it in the circumstances that I am in, which makes my obligations to him so much the greater. (*Conway Letters*, 534)

Although she acknowledges the value of van Helmont's medicines, she emphasizes the relief she finds in his patient company, a suggestion that his aid consisted as much if not more in a shared mode of being than in medicinal treatment.

That shared mode of being involved van Helmont's contributions to Conway's interest in both Kabbalistic literature and later in Quakerism, contributions that are well known.[33]   Central to both interests, however, was his attitude toward pain, which is illustrated in an autobiographical narrative he told regarding his own experience of pain. During one of his travels, a large tree fell on him, crushing the bones in his shoulder. He was given time to contemplate his pain because his doctors were slow and frustratingly ineffective. As the broken bones joggled against one another, inflicting intense pain, his frustration gave way to surprise as he observed that his pain had a transmutative quality; because the pain was, as he put it, "nothing less, but my owne life, excited or inflamed for my good....I began to love my pain."[34] Physical pain not only signaled his life; it also signaled a force acting "for [his]

---

[33] Allison P. Coudert's work on van Helmont is definitive. See Allison P. Coudert, "Henry More, the Kabbalah, and the Quakers," in *Philosophy, Science, and Religion in England, 1640–1700*, ed. Richard Kroll, Richard Ashcraft, Perez Zagorin, 31–67; Allison Coudert, *The Impact of the Kabballah in the Seventeenth-Century: The Life and Thought of Francis Mercury van Helmont, 1614–1698* (Leiden: Brill, 1999); Coudert's "A Cambridge Platonist's Kabbalist Nightmare," *Journal of the History of Ideas*, 35 (1975): 633–52; Marjorie Nicolson, "The Real Scholar Gypsy," *Yale Review* 18 (1929): 353–54;

[34] As cited in Nicolson, *The Conway Letters*, 315. The manuscript cited by Nicolson is in the British Library, Sloane 530.

good." Influenced by his father's Paracelsian mysticism and by his own research into Kabbalistic writings, van Helmont devised a meliorative theory of suffering through which the individual self could experience Christ. Allison Coudert has noted the link between pain, alchemy, and Christianity:

> There was no greater symbol of the value of pain and suffering than the figure of Christ crucified and resurrected. Alchemists incorporated this idea into their art, making it an axiom that the physical suffering and death of the compounds they employed were the essential steps in their purification and transmutation. With this alchemical background and unorthodox Christian beliefs van Helmont came to view pain as a necessary but transitory state in the drama of universal salvation. In convincing Lady Conway of this truth he gave her help as a philosopher where he could not give it as a physician.[35]

Van Helmont's experience of pain helps to explain why he was drawn both to Quakerism and to Kabbalistic learning, both of which focused on pain as an agent of spiritual refinement. His theory of pain was informed, as Allison Coudert's recent biography notes, by the disciples of the sixteenth-century Kabbalist, Isaac Luria.[36] Through pain and suffering, Luria's disciples argued, individuals could eventually be restored to their pre-fallen state through a series of reincarnations. Coudert observes that van Helmont extended Luria's theory of the transmigration of souls to argue that "eventually the whole creation, including the very dirt of the earth, will be raised up and perfected."[37] Pain has, in van Helmont's theory, the same transmutative force described in the narrative of his breaking his shoulder, cited above: it becomes the

---

[35] Coudert, "Henry More," 38.

[36] See Coudert, *The Impact of the Kabballah.* See also Coudert's "A Cambridge Platonist's Kabbalist Nightmare," 633–52.

[37] Coudert, *The Impact of the Kabbalah*, 196–97.

agent of spiritual millenarianism. Scholars have noted the logic of Conway's turning to spiritual strategies for coping with pain after it became clear that no physical cure could be found, but little discussion has focused on why Conway abandoned More's rational theology to take up van Helmont's mystical Christianity.[38] Conway's interest in Kabbalistic theory reveals her interest in using theological issues to subvert contemporary philosophical systems. She recognized that the intense focus on the suffering self offered by Kabbalistic theory served as a critique of the new science's philosophical systems. Where the new science had, in Conway's mind, alienated the self from the world by failing to account for pain, Kabbalistic theory returned the self to the center of philosophical attention.

In David Morris's terms, van Helmont's theory of pain transferred the focus on pain from "sensation" to "perception":

Sensations, like heat and cold, require little more than a rudimentary, functioning nervous system. A salamander or a juice bug can experience sensation. Perceptions, by contrast, require minds and emotions as well as nerves. When understanding pain as perception, we are implicitly challenging the deeply entrenched mechanistic tradition in medicine that treats us as divided into

---

[38] Sarah Hutton notes that "in light of what we know of the unrelievable physical pain from which Anne Conway suffered most of her life...it is quite understandable that she could not accept a dualistic separation of mind from body.: See also Hutton's "Anne Conway, Margaret Cavendish and Seventeenth-Century Scientific Thought" in *Women, Science and Medicine 1500-1700*, ed. Lynette Hunter and Sarah Hutton (Stroud, Gloucestershire: Sutton, 1997), 229. Elsewhere, Hutton similarly notes that "Conway's experience of pain was a significant determinant in the development of her mature philosophy." See "Of Physic and Philosophy," 229. Finally, an overview of Conway's relationship with van Helmont is provided in Hutton's *Ancient Wisdom and Modern Philosophy: Anne Conway, F.M. van Helmont and the Seventeenth-century Dutch Interchange of Ideas* (Utrecht: Utrecht University Press, 1994).

Allison Coudert and Taylor Corse note that Conway's focus on the problem of pain reveals "both how perceptive [she] was and how useful biography can be in interpreting the work of a philosopher." See their introduction to their translation of *The Principles of the Most Ancient and Modern Philosophy* (Cambridge: Cambridge University Press, 1996), xvi.

separate and uncommunicating blocks called body and mind.[39]

It would become Conway's project in *Principles of the Most Ancient and Modern Philosophy* to argue against theories of matter that posited "uncommunicating blocks called body and mind." Van Helmont's focus on the value of perceiving pain as opposed to sensing pain helps to explain his influence on Conway, who by the early 1660s had come to accept her own pain not as a temporary disruption of her life but as its defining characteristic. Her debt to van Helmont and to Kabbalistic thought is evident in *Principles,* in which she uses the problem of pain to refute both Hobbesian materialism and Cartesian dualism. In refuting the latter, she was also refuting More's dualistic system.

Pain presented and in fact continues to present difficulties for materialists in the tradition of Thomas Hobbes, who reduce the world to pure matter to be accounted for through the laws of physics. The pain of a headache, like the notion of a thinking self, resists materialist accounts, since both can be perceived without necessarily identifying a material basis. How, furthermore, can inert matter feel? Pain poses no less difficulty for Cartesian dualists, who posit two categories of reality: one body, the other spirit. Dualists face the problem of having to account for the interaction between body and spirit, a mystery for which no entirely satisfactory account exists. As Joseph Glanvill famously put it:

> How the purer Spirit is united to this clod, is a knot too hard for fallen Humanity to unty....How should a thought be united to a marble-statue, or a sun-beam to a lump of clay! The freezing of the words in the air in the northern climes, is as conceivable, as this strange

---

[39] Morris, *The Culture of Pain*, 75.

union....And to hang weights on the wings of the winde seems far more intelligible.[40]

Neither Hobbesian materialism nor Cartesian dualism presented altogether satisfactory explanations for the phenomenon of pain, an observation Conway incorporates into her own monistic system, which posited that everything was spirit and that the distinction between body and spirit was only one "of mode, not essence."[41] As she put the problem,

> Why does the spirit or soul suffer so with bodily pain? For if when united to the body it has no corporeality or bodily nature, why is it wounded or grieved when the body is wounded, whose nature is so different? For since the soul can so easily penetrate the body, how can any corporeal thing hurt it? If one says that only the body feels pain but not the soul, this contradicts the principle of those who affirm that the body has no life or perception. But if one admits that the soul is of one nature and substance with the body, although it surpasses the body by many degrees of life and spirituality, just as it does in swiftness and penetrability and various other perfections, then all the above mentioned difficulties vanish; and one may easily understand how the soul and body are united together and how the soul moves the body and suffers with and through it. (*Principles*, 58)

Conway places pain at the center of her system, grounding her philosophic system in the experience of the self.

She had earlier tried to persuade Henry More privately of the error of his own dualism through their exchange of stories regarding supernatural events. More, like his friend Joseph

---

[40] Joseph Glanvill, *The Vanity of Dogmatizing* (London, 1661), 20.
[41] Conway, *The Principles of the Most Ancient and Modern Philosophy*, 41. All subsequent citations are to this edition and will be made within the text. I have also relied on the introduction to Peter Loptson's edition (The Hague: Martinus Nijhoff, 1982).

Glanvill, took an interest in arguments positing the existence of witches because such arguments served to preserve a disappearing spirit world. Conway, however, used those same arguments to call into question More's dualism. If, she asked More in 1671, the souls of witches leave their bodies to meet at nocturnal conventicles, why is it that the souls of good people do not similarly leave their bodies to visit their distant friends? (*Conway Letters*, 347). Alluding directly to the pain of a cherished and intense long-distance relationship, the question was intentionally provocative. More's response to her question seems utterly unsatisfactory: "There may be some damage or diminution," he returned, "done to the vitall union of that soul and body, which may make the life of the party less comfortable." "But," he quickly added, perhaps sensing Conway's implicit critique of his own system, "those are curiosityes about which we need not be at all solicitous" (*Conway Letters*, 347).[42]

It is perhaps not coincidental that Conway was not the first to question More's dualism. Margaret Cavendish had earlier presented a critique of More's rational theology in *Philosophical Letters* (1664), which devotes equal attention to Descartes, Hobbes, More, and J.B. van Helmont, Francis Mercury's father. Like Conway, Cavendish posited a vitalistic monistic universe in which everything is sentient. More had defended his dualism by arguing that some matter–a stone, for instance–is obviously inanimate. Were stones animate, they would surely move out of the way

---

[42] In their lucid introduction to Conway's *Principles,* Allison Coudert and Taylor Corse note the parallel between Conway's critique of More's dualism and the Palatine Princess Elizabeth's critique of Descartes's dualism:

> In raising this question, Lady Conway echoed an objection made by one of Descartes' correspondents, the Palatine Princess Elizabeth, who was the grand-daughter of James I of England and eldest daughter of the unfortunate "Winter" King and Queen of Bohemia.... In one of her letters to Descartes she posed the same question Lady Conway was later to raise: Why, if the mind is utterly unlike the body is it so troubled and worried by physical feelings, sensations, and emotions? Descartes never answered the question. He simply advised Elizabeth to spend only a few days a year on metaphysical matters, something he certainly never suggested to any of his male correspondents who queried aspects of his philosophy. See "Introduction," xvi–xvii.

when carts approached. Cavendish scoffed at this account by
noting that a stone's inability to avoid pain does not prove that it
is inanimate; humans, who are certainly animate, are not much
more successful than stones at avoiding pain:

> Not any human Creature, which is accounted to have the
> perfectest sense and reason, is able always to avoid what
> is hurtful or painful, for it is subject to it by Nature: Nay,
> the Immaterial Soul it self, according to your Author,
> cannot by her self-contracting faculty withdraw her self
> from pain. (*Philosophical Letters*, 193)

Stones have a sort of sense and reason, she argued, though these
operate differently in them than they do in humans. Cavendish's
vitalism, like Conway's, resulted in a universe in which everything
is potentially vulnerable to pain.[43] And like Conway, Cavendish
was unwilling to segregate the human world from the natural
world as More did in claiming that reason made man "the flower
and chief of all products of nature."[44] For both women, the human
world was not above nature but within it.

---

[43] In part, Cavendish's critique of More focuses on his theological speculation. Because God
was, for Cavendish, utterly unknowable, she argued that such speculation as More's was
both idle and potentially disruptive to the polity. She relegated his volumes to the category
of "disputes."

> In things Divine, Disputes do rather weaken Faith, then prove Truth, and breed
> several strange opinions....which is the cause of so many schismes, sects, and
> divisions in Religion. (221)

Though Conway and Cavendish share superficial similarities, they are very different
thinkers. The far more pessimistic Cavendish saw the problem of pain as an indication of
God's remoteness from the natural world and of his unknowability. "Nature," she allowed,
could "be known in parts, [but] God being incomprehensible, his Essence can by no wayes
or means be naturally known" (*Philosophical Letters*, 140). Conway's millenial vision of
eventual perfection of all things seems by contrast optimistic.

For another comparison of Conway's and Cavendish's systems, see Peter Loptson's
Introduction to *The Principles of the Most Ancient and Modern Philosophy*, 67n64.

[44] Cavendish cites this phrase to take issue with it in *Philosophical Letters*, 147.

Cavendish sent Henry More a personal copy of *Philosophical Letters* in 1665. His reaction is recorded in a letter he immediately wrote to Conway:

> My Lady of Newcastle has sent two more Folios of hers to furnish my study, the one of poems, the other which is far the bigger, of letters wherein I am concern'd, above 30 of those letters being intended for a confutation of sundry passages in my writings. I wish your Ladiship were rid of your headache and paines, though it were no exchange for those of answering this great Philosopher. (*Conway Letters*, 237)

His condescension makes clear that he dismissed Cavendish's critique just as peremptorily as he would overlook Conway's similar critique of his dualism six years later.

But Conway, who was in many ways different from Cavendish, may well have been more receptive than More to Cavendish's critique of his dualism and particularly to Cavendish's attention to the problem of pain. When she composed *Principles*, shortly after van Helmont's arrival at Ragley, she, too, opposed More's dualism, though she did so by appropriating van Helmont's Kabbalistic theories to posit a monistic, vitalist system in which everything is spirit, and in which pain functions as a refining force.[45] Her account of pain explains her attraction to the

---

[45] Peter Loptson has suggested that the date of composition of Conway's *Principles of the Most Ancient and Modern Philosophy* is contingent on the arrival at Ragley of copies of the *Kabbala denudata*; he therefore dates the composition of *Principles* as roughly "the two-year period from early 1677 to Conway's death in February 1679–early rather than late in this period, as she [would] have been too ill to write in the last six months...of her life" (17). Loptson concedes that More received advance chapters of the *Kabbala* from "at least 1672 onwards" (18). In choosing this date, he departs from Marjorie Nicolson, who dates the composition of *Principles* to sometime during 1671–1674, arguing that "the absence of any influence of Quakerism shows clearly that it [the notebook containing *Principles*] was laid aside before her interest in that movement began" (*Conway Letters*, 453). Conway's political involvement on behalf of the Quakers after 1675 and her increasing interest in Quakerism itself after 1677 would seem to have taken time away from natural philosophy, and thus Nicolson's conjecture seems more probable than Loptson's. See Loptson, *The Principles of the Most Ancient and Modern Philosophy*.

Quakers, who similarly found in pain evidence of Christ's immediate work within. For Conway, pain effected change, aiding darker, more corporeal spirit to be refined into lighter, more spiritual spirit.

> [A]ll pain and torment stimulates the life or spirit existing in everything which suffers. As we see from constant experience and as reason teaches us, this must necessarily happen because through pain and suffering whatever grossness or crassness is contracted by the spirit or body is diminished; and so the spirit imprisoned in such grossness or crassness is set free and becomes more spiritual and, consequently, more active and effective through pain. (*Principles*, 43)

By turning to the metaphor of hard-heartedness and arguing for its literal meaning, Conway suggests that pain not only reliably effects virtue; it also indicates virtue: "the heart or spirit of a wicked man is called hard and stony because his spirit has indeed real hardness in it...On the other hand, the spirit of a good person is soft and tender" (*Principles*, 44). The implied corollary in this anatomy of virtue is that a good person simply feels pain more readily than does a wicked person. Pain, then, becomes both a source and an indicator of virtue:

> Those who are dead in their sins lack this sense of the hardness or softness of good or bad spirits, and for this reason they regard these phrases as merely metaphors, when, in fact, they have a real and proper meaning without any figurative sense. (*Principles*, 44)

To experience pain was thus to perceive and to exercise one's own virtue.[46]

---

[46] For a discussion of how Conway provides a fuller (though not entirely satisfactory) account of interaction between varying degrees of spiritual parts than Descartes does between spirit and matter, see Jane Duran, "Anne Viscountess Conway: A Seventeenth

Under Conway's system, then, the body becomes a microcosm; by turning inward, by perceiving pain or suffering, one could be transformed and experience Christ, the embodiment of the redemptive power of suffering.   Conway gives pain the transmutative power that van Helmont attributed to it. It was pain that transformed the hard-hearted into the tender-hearted. Paradoxically, it was pain that allowed one to experience the refreshment of Christ or the inner light.   Where Hobbesian materialism or Cartesian dualism had essentially alienated the self from its world, Conway's vitalism offered a way of reconciling self and world. She devoted herself to Kabbalistic thought out of an optimistic "millenial vision of a future time when every particle of matter would be restored to its pristine spiritual condition."[47] For Allison Coudert, such optimism was characteristic of the Cambridge Platonists:

> A hallmark of the Cambridge Platonists was their rejection of the Augustinian and Calvinist emphasis on man's depraved and fallen nature and the reassertion of the renaissance theme of the dignity of man. The optimistic and inherently Pelagian outlook of the Cambridge Platonists was reflected in their preference for the writings of the Greek philosophers and Greek Church fathers over those of Augustine and later Churchmen. As Patrides has said, "The deification of man is one of the most thoroughly Greek ideas espoused by the Cambridge Platonists."[48]

Coudert concludes by claiming that despite More's early interest in the Kabbalah and Quakerism, he could not, finally "accept the optimistic, radical Pelagianism inherent in the occult philosophies

Century Rationalist," *Hypatia* 4:1 (Spring 1989): 64–79.
[47] Coudert, "Henry More, the Kabbalah, and the Quakers," 56.
[48] Coudert, "Henry More, the Kabbalah, and the Quakers," 48. Coudert quotes C. A. Patrides, *The Cambridge Platonists* (Cambridge: Cambridge University Press, 1978), 19.

to which he was otherwise so obviously drawn."[49] Conway clearly was drawn to the optimism that More finally rejected. Her insistence that the perception of pain was itself evidence of Christ's work was simply too positive for More, who distrusted both dogmatism and certainty.

Conway's conversion to Quakerism around 1677 distressed More, who distrusted any sign of sectarian enthusiasm: he "receiv'd the Account of it with Tears, and labour'd, all that a Faithful Friend could do, to set her right, as to her Judgement in these Matters" (*Life*, 120). His efforts, however, failed. She preferred Quakers in her household, she explained in a letter to him, because as "lovers of quiett and retirement," they more usefully than others "fit the Circumstances I am in, that cannot endure any Noise" (*Conway Letters*, 422). Her final refutation of his theological and philosophical structure came when she pointedly explained that she preferred the refreshment she found with the Quakers to all other comforts:

> They have been and are a suffering people and are taught from the consolation [that] has been experimentally felt by them under their great tryals to administer comfort upon occasion to others in great distresse....The weight of my affliction lies so very heavy upon me, that it is incredible how very seldom I can endure anyone in my chamber, but I find them so still, and very serious, that the company of such of them as I have hitherto seene, will be acceptable to me, as long as I am capable of enjoying any; the particular acquaintance with such living examples of great patience under sundry heavy exercises, both of bodily sicknesse and other calamitys (as some of them have related to me) I find begetts a more lively fayth and uninterrupted desire of approaching to such a behaviour in like exigencyes, then the most

---

[49] Coudert, "Henry More, the Kabbalah, and the Quakers," 59.

learned and Rhetorical discourses of resignation can doe.
(*Conway Letters*, 421–22)

More had for twenty-six years written her devoted letters bursting with "learned and rhetorical discourses of resignation." Conway's preference for "experimentally felt" consolation over such "learned discourses" signaled a decisive philosophical departure from his very way of being. She turned inward to the mystical and transformative experience of the inner light and away from the rhetorically sophisticated theological abstractions of More's rational theology. Her rejection of his rational theology accompanied her rejection of his dualism. If "the gulf that existed between matter and mind, or matter and spirit, in More's philosophy paralleled the gulf between man and God," the absence of that gulf in Conway's monism similarly paralleled her belief in the union of the human and the divine.[50] Her pain was for her evidence of Christ working within her as the agent of redemptive agony.

Like the banished Duke in William Shakespeare's *As You Like It*, both Conway and More found in pain a "counsellor/ that feelingly persuade[s] me what I am" (II.i 10–11). Both understood what David Morris observes: "Pain...is far more than simply or exclusively a...transaction of the nervous system. Further, pain is not always an unmitigated disaster. We willingly, if grudgingly, accept pain that accompanies growth or achievement...[pain] makes us who we are" (*The Culture of Pain*, 20). Yet Conway and More differed, finally, in their respective responses to pain. Willing to acknowledge the utility of "the Furnace of Affliction," More could not, however, accept a theory of pain that appeared to remove doubt and spiritual struggle from religious experience. Instead, he sought explanatory discourses for pain by devoting himself to a discursive mode and producing a prolific number of volumes that returned repeatedly to the theme of Christian resignation. Distrustful of enthusiasm, More could at times insist that God himself was purely discursive and rational: "God doth

---

[50] Coudert, "Henry More, the Kabbalah, and the Quakers," 56.

not ride me as a horse, and guide me I know not whither myself;
but converseth with me as a Friend; and speaks to me in such a
Dialect as I understand fully, & can make others understand, that
have not made shipwrack of the faculties that God hath given
them, by superstition or sensuality."[51]   Consumed by pain,
Conway supplanted More's discursive and distant God with a God
with whom she could be united through her daily experience of
pain.

More spent much of the summer of 1677 at Ragley Hall, but
as Marjorie Nicolson notes, "Lady Conway for the first time had
little real need of him" (Conway Letters, 434). Her husband noted
that though Quakers had free access to his wife's room, "Dr. More,
though he was in the house all last summer, did not see her above
twice or thrice" (Conway Letters, 434). And though More may have
continued to write to Conway, his letters do not survive. In their
place are letters from Quakers like Lillias Skene, urging Conway to

> f[aint] not nor be weary under thy personal trayels bot
> feel tribulation working patience and l[et] patience have
> its perfect work. My desirs are that more and more that
> eye may be opened in thee that looks beyond the things
> that are seen, and in the living sence of the invisable
> glorie of the kingdom of god thou may live above the
> desir of temporarie satisfactions, crying secretly in thy
> heart that by all thy present sufferings th[y] iniquetis
> may be perched away and sensteev netur be crusified, in
> the death of which shal he com to regn whos right it is,

---

[51] Henry More, *Enthusiasmus Triumphatus, Or, a Discourse of the Nature, Causes, Kinds, and Cure, of Enthusiasme* (London, 1656), 182–83. Richard Ward quotes this passage as being fully characteristic of More's attitude toward God. See Ward, *Life*, 38. *Enthusiasmus Triumphatus* was admittedly an argument against enthusiasm, which together with atheism More considered to be twin threats to Christianity and to the Church of England. Elsewhere, More himself admitted to having "a Natural touch of Enthusiasme in his Complexion; but such as (he thanks God) was ever governable enough; and which he had found at length perfectly Subduable" (Ward 35). As Robert Crocker notes, More's "enthusiast opponents in effect held up for public appraisal an uncomfortably distorted mirror to his own aspirations and ideas" (150). See Robert Crocker, "Mysticism and Enthusiasm in Henry More" in Hutton, *Henry More*.

and the conforter whom no man can tak from thee will
dwel in thee and abyd with thee for ever.   (*Conway
Letters*, 439)

Skene's confidence in pain's "perfect work" in "p[uring] away"
iniquities and "crusify[ing] sensteev nature" reflects her faith that
pain alone could effect change and unify one with God.

Though More no doubt would have been troubled by such a
precise equation between pain and redemption, Conway found in
it a needed "encounter with meaning." Like van Helmont's theory
of pain, Skene's equation of pain with the redemptive power of
Christ helped to transform the experience of pain from mere
sensation to transformative perception. Faced with constant and
debilitating pain, Conway's language ran dry. At her death, she
left the slender manuscript of *Principles* in a notebook among her
papers with no instructions for publication. Were it not for the
efforts of van Helmont and More to see it through publication, we
might not have it today. Even in death, Conway shunned
discourse, asking to be buried without "the Rites and Ceremonyes
of the so-called Church of England" (*Conway Letters*, 481). Her
only epitaph is the two-word phrase "Quaker Lady" inscribed on
her lead coffin. To make sense of her pain, she found it necessary
to abandon the ways of her voluble friend whose profound loyalty
must have rendered more painful the futility of his discursive
endeavors to help her.

THE CONSUMPTION OF MEAT
IN AN
AGE OF MATERIALISM
Materialism, Vitalism,
and Some
Enlightenment Debates over
National Stereotypes

George Sebastian Rousseau
Oxford University

I relate the substance of a fable about a part of nature and its representation in a bygone era. The natural substance is *meat*: its existence, cultivation, preparation, consumption by humans, and—of course—its articulations, verbalizations, narratives, mythologies, religious associations, iconographies. What is meat apart from these religious and secular representations? Meat barely exists otherwise, except to eat. Ingest meat silently—*un*mentally—and it loses all meaning except the nutritive protein one. Digest it accompanied by verbalization and narrative (mentalization in words), even surrounded by atmosphere and music (merely ponder its role in eighteenth-century Italian opera), and its bearings become altogether different, as the whole of human and religious history demonstrates, not merely in Judæ-

Christian history but Islamic, Hindu, Oriental, African, tribal and all other human histories. I relate only a tiny piece of the story here, chronologically located in the heartland of the eighteenth century but outreaching both earlier to the seventeenth century and later to the nineteenth. Hovering everywhere, but unarticulated, around the margins of my account are the sustained eighteenth-century debates over materialism and vitalism: meat as privileged type of matter, meat as phenomenon *other* than matter forever existing as representation and mythology far more than mere material substance. The narrative below will be sufficiently intelligible as fable. I leave it to readers to extract the moral from my historical narrative, particularly its value for, and relation to, the scientific materialism-vitalism controversies of the European Enlightenment. Only this moral will be ambiguous. How could we expect another profile for the substance that formed the basis for transubstantiation and the Eucharist in Christian history? How could "meat" have been less problematical? Especially when we recollect that there is no blood apart from meat.

I: Stereotypes of the Fierce English: The Raw and the Cooked

A remarkable passage near the beginning of La Mettrie's *Man a Machine* (Leiden, 1748, in French; the first English translation one year later, 1749) embeds the *topos* of the raw and the cooked that will run its course to the twentieth century, especially to Levi-Straus and the anthropologists.[1] The context is La Mettrie's

---

[1] See Claude Levi-Straus, *The Raw and the Cooked* (London: Penguin, 1992, trans. John and Doreen Weightman). The vast literature on cannibalism—literal and metaphoric—has also shaped the contexts; see, for example, Frank Lestringant, *Cannibals: The Discovery and Representation of the Cannibal from Columbus to Jules Verne*, tr. Rosemary Morris (Berkeley and Los Angeles: University of California Press, 1997). Another discourse about meat in the early modern world is also important; see Giovanni Rebora, "Meat," in *Culture of the Fork: A Brief History of Food in Europe* (New York: Columbia University Press, 2001), 43-51, who claims that the price of beef was cheap relative to the rest of food despite the high cost of fine cuts of meat. More general works have also demonstrated what food and its eating has meant;

introductory argument about the body's influence on the "soul" for practical proofs of which he offers the way coffee impinges on the imagination to produce headaches, and the consequences of a good supper.[2] La Mettrie then continues to claim that the British eat their meat raw to remain fierce:

> Raw meat gives a fierceness to animals; and man would become fierce by the same nourishment. This is so true, that the *English*, who eat not their meat so well roasted or boiled as we, but red and bloody, seem to partake of this fierceness more or less, which arises in part from such food, and from other causes, which nothing but education can render ineffectual. This fierceness produces in the soul pride, hatred, contempt of other nations, indocility, and other bad qualities that deprave man's character, just as gross phlegmatic meat causes a heavy, cloudy spirit, whose favourite attributes are idleness and indolence."[3]

see anthropological studies such as Peter Farb and George Armelagos, *Consuming Passions: The Anthropology of Eating* (Boston: Houghton Mifflin, 1980). I am grateful to the Leverhulme Trust for funding some of the research which made this study possible.

[2] See the classical study of A. Vartanian, *"L'Homme Machine": A Study in the Origins of an Idea* (Princeton: Princeton University Press, 1960), as well as Ann Thomson's *Materialism and Society in the Mid-Eighteenth Century: La Mettrie's "Discours préliminaire"* (London: Macmillan, 1981) and her *Julian Offray de La Mettrie, Machine Man and Other Writings—Cambridge Texts in the History of Philosophy* (Cambridge: Cambridge University Press, 1996).

[3] *L'homme-machine* was first published in Leiden in 1748 by Elie Luzac *fils* and sentenced to be burnt. It was republished in the first editon of Julien Offray de La Mettrie, *Œuvres philosophiques*, probably in Berlin together with the *Traite de l'ame*, as well as an introductory *Discours preliminaire (l'histoire naturelle de l'ame), l'abrege des systemes pour faciliter l'intelligence, l'homme plante, les animaux plus que machines, le systeme d'epicure*. See Francine Marcovits (ed.), *Julien Offray de La Mettrie, Œuvres philosophiques* (Paris: Fayard, 1987, 2 vols.). This passage is found in the first English translation (translator unknown): *Man a Machine, translated from the French* (Dublin, 1749), 13. The passage's context exclaims the wonderful power of repasts: "What a vast power there is in a repast! Joy revives in a disconsolate heart; it is transfused into the souls of all the guests, who express it by amiable conversation, or music. The hypochondriac mortal is overpowered with it; and the lumpish pedant is unfit for the entertainment."

What can La Mettrie mean by such cause and effect? To what does he refer in this didactic pronouncement about English meat, so glib it seems almost to be an accepted conceit? Surely the English cooked their meat and abundant evidence exists that they did. Why then this mythology, widespread in the epoch, and how can any Anglomania—the love of the English—have developed in the face of such barbarism? For this is ferocity of a *new* type: the fierce as code word for pride, sin, transgression, all exceeding the merely crude and raw carnivore apart from the vegetable kingdom La Mettrie himself had described in *L'homme plante* in the second volume of his philosophical works.[4] By ferocity La Mettrie repeats the Voltairean claim of the 1720s about the raw meat-eating habits of the Brits.[5] The paradigm is simple enough and may antedate Voltaire: societies which cook food are civilized and refined, in part because the fire used to cook it, as well as the servants and other domestic help needed to prepare it, are indicative of the consumer's status within an already developed civilization. These were crucial distinctions in an epoch when national stereotypes assumed new levels of importance, especially among the premier nations—England and France—then vying for European supremacy. The Portugese and Spaniards had already lost their far-flung empires. The Dutch, nearly having lost theirs, were dwindling into the second-rate power they would become after they had traded New Amsterdam for a tiny nutmeg island in the East Indies. The Germans and Italians had never amassed an empire to lose, were not yet coherent nations, and—to gaze further back—a very long time had elapsed since the Homeric epics recorded the routine roasting of meat on spits and since Marco Polo noted in his thirteenth-century travel journals, when in the valleys of Szechuan and the mountains of Yunnan in western

---

[4] La Mettrie, *Œuvres philosophiques* (Berlin: Charles Tutot, 1796, 3 vols.); *L'homme-plante*, 2: 49-75, especially chap. 3.
[5] La Mettrie was aware of Voltaire's comments; see La Mettrie, *Œuvres philosophiques* (London: Jean Nourse, 1751), 209, note 36.

China, that both gentry and poor "eat their meat raw."[6] Today we are surprised if such symbolic distinctions were applied to countries and peoples, although stories about Italian cuisine were already filtering back to England in travel books such as Tobias Smollett's.[7] In eighteenth-century France and England, however, the consumption of food was another matter.

Even so, other interpretations should be considered, such as the one investing in the pragmatics: the physical act of chewing, masticating, digesting the raw versus cooked food. For not all lies in preparation but in mastication. Imagine the difference in that era—the English Restoration after Charles II and early eighteenth century in the period of Anne and the first Georges—when huge segments of the population in both countries were eating themselves into the grave. Even without teeth, liquids were drunk, swallowed down the throat directly into the stomach. It made a difference whether cold or hot liquid and analogies can be drawn between raw and cooked, and cold and hot; but consider at them more closely and liquids appear as another order. Likewise sauces, pastes, and gravies, then proliferating. The range of foods requiring proficient teeth—chewing, mastication—was another matter. In that era of taxonomic upheaval and materialistic-vitalistic controversy, when the Buffons and Linnaeuses were reconfiguring the terraqueous world according to their new classifications, food was also classified according to the potable and the chewable, the lenitive (or palliative) and the odontological. Cookbooks and didactic treatises confirmed the differences in their taxonomies. It was not merely a matter of the raw and the cooked but much more (along this line) of the odontological and non-denticular.[8]

Recent scholars have noticed to what degree theirs was a culture of the mouth: vocality and articulation in the Republic of

---

[6] Cited in Reay Tannahill, *Food in History* (New York: 1988, rev. ed.), 137; see also his *Flesh and Blood: A History of the Cannibal Complex* (New York: Stein and Day, 1975).

[7] *Travels through France and Italy* (London, 1766) where he comments on Italian food.

[8] A fascinating discussion of odontological imagery from ancient mythology to modern European fiction is found in Theodore Ziolkowski, "The Telltale Teeth: Psychodontia to Sociodontia," *PMLA* 91 (1976): 9-22.

Letters; the new æsthetics of the voice in the widely developing opera; kissing and other oral arts in a cosmopolitan Georgian culture quickly becoming eroticized for commercial purposes. When writers as diverse as Montesquieu in the *Lettres Persanne* and Goldsmith in the *Citizen of the World* contrasted the European kiss with the Oriental kiss, the Enlightenment mouth seems to have run its full geographical gamut.[9] Yet food, especially meat and the customs attached to meat-eating, in France and Britain, has been somewhat overlooked and undervalued as the clue to a sign system about national stereotype and nationhood.[10] It is not that meat and mastication are aberrant topics (perhaps as sex was forty years ago in academia), or without literary and philosophical consequence, but that the paradigm's end product can be viewed as historically rather meagre: the French eat cooked meat (so the myth went) and the British ate it raw; the French are refined (in this mythology), the English crude. What more is there to say, even in an era marked by exquisitely differentiated national stereotypes from Defoe to Goldsmith, Voltaire to Madame de Stael and the Schlegel brothers?

But is La Mettrie's paradigm actually so trifling? If food is a key to national temperaments—at least a sign of the main composite differences—then one learns much about countries, Eastern and Western, by consulting their food: its production and growth, its preparation and serving arrangements, and, of course, its cookery. One might say the same of teeth—that you can interpret countries denticularly (materialism again: is everything material?); but a semiotics of mastication seems to be more empirically based than the raw and cooked will be. Perhaps it is best to document this not-so-minor point briefly, without becoming diverted, before moving on to the raw and cooked.

[9] Walter Ong's important study of the modern culture of orality in *Orality and Literacy: the Technologizing of the Word* (London: Routledge, 1988), has been unhelpful here.
[10] Food as a category or topic is not mentioned in Cedric Barfoot's large and useful anthology of national stereotypes; see *Beyond Puig's Tour: National and Ethnic Stereotyping in Theory and Literary Practice*, ed. C. C. Barfoot (Amsterdam: Rodopi, 1997).

Teeth in the eighteenth century were not what they are now. Numerous are the accounts of the sorry state of their condition: how few teeth were actually *in* mouths, how inferior their wooden replacements were, the excruciating pain of mastication, the swollen and mucilaginous gums which—one textbook about teeth claimed—bled so profusely the leech was sent away the next day when he came to bleed. Eighteenth-century painting may idealize the teeth of its subjects, but pictorial satires provided a more realistic picture of the putrid state of the mouths in that age before Chanel mouthwashes and Colgate flourides. One author vividly described the cavities between the teeth: not clean, high-flossed, empty channels, but sewers of muck, decayed food, and microscopic worms. The tooth was inset to this sea of accumulating filth; there it decayed, atrophied, and died, which is why the tooth-puller was such a necessary, if omnipresent figure, as painters since the Renaissance had shown. Wooden dentures replaced it, but to masticate with them, especially raw food, was a very different matter from the natural enamel-biting teeth God-given at birth. These wooden dentures were massive apparatuses, several surviving examples of which are displayed in the National Library of Austria in Vienna; so large, so thick, so imposing, so piercing in the way they jutted into the material gums and inevitably bled them with every bite made by mastication that most users refrained inserting them. Moreover, they were expensive, unafford-able to all except the well-heeled, and were bought by special order with an average wait-period of over one year.

Such a brief synopsis of mastication does not, of course, do justice to this (depending on your viewpoint) vivid or grotesque subject. But without mastication the earlier matter of raw and cooked and the consumption of meat loses its historical valence. For meat was then a most costly food when served in good cuts save for rarities such as pineapples which had to be grown in a hothouse (you will remember that Smollett's novel *The Adventures of Peregrine Pickle* opens with the pregnant Mrs. Pickle craving a pineapple; Gamaliel, her husband, wanders the countryside to find her one; when Gamaliel locates it, she cannot have it because of its exorbitant price). Equally expensive were imported exotic delicacies which only the rich could afford. Other than these, meat economically headed the list, there being less consumption of fish then than there is today (and to this day the British remain highly suspicious of fish and, with the exception of fish and chips, eat much less fish than any other European country.) You can tell much about a nation—past and present—by reading the history of its meat-eating customs, even if most medical treatises warned the consumer about the vermin its caused in the gut and the worm-eating animals it produced in the bowels. The eighteenth-century Enlightenment was no Era of Vegetarianism despite its quondam devotees from Thomas Tryon in the 1690s and George Cheyne in the 1720s to scattered utopians cultivating their allotments dotting the long century leading down to vegetarian Shelley. As a culture of meat-eaters the Enlightenment view was transformative and intensely aware that meat-eating existed in a class apart from the consumption of other foods. This does not include the emotional effects. We must not forget that however dead the old humoral pathology was, it was still being applied to remedy the temperaments, and where food played a major role, as when surgical apprentice Roderick Random in Smollett's novel accommodates with apothecary Lavement (suitably named), whose adolescent daughter and ruddy wife are right lusty wenches whose behaviors have deranged the apothecary father. Roderick's medical advice is: remove meat from her rumpy diet, which will cool her down.[11] One Smollett annotator speculates that meat should be viewed in the context then of body warmth, that is, temperature. This may be so, but there is no contemporary evidence that the British Augustans thought that they would be kept warmer by consuming raw meat than cooked.

There had always been concern about meat-eating in Europe, especially for its pejorative effects on the anatomical organs, even before the doctrines of correspondence (the notion that a walnut

---

[11] Smollett, *The Adventures of Roderick Random* (London, 1748), ch. 19, the passage about Lavement's repeated attempts "more than once to introduce a vegetable diet into his family."

looks like a brain—therefore if the brain is troubled, eat walnuts for palliative restoration). The medieval Cathars apparently avoided meat altogether, and other religious orders followed their practice. Even Erasmus explained how problematic its preparation was. The issue was not the avoidance from flesh during Lent, but abstinence altogether: an inverse of cannibalism. Milk, not meat, was the medicinal wonder-drug of the seventeenth century, its virtues classified according to its animal provider: camels to goats, asses to cows, and especially human milk. Milk above all was prescribed when medical consumption ravaged societies; not what we call phthisis or tuberculosis but a wasting condition *not* limited to the pulmonary cavity (here it is crucial to remember that for centuries before the eighteenth, consumption, or the "tabes," was a generalized, gender-free, rather democratic condition which ravaged human bodies without respect of sex, age, or class—the class insertion was a nineteenth century addition after it was exclusively moved to the pulmonary cavity.) Thomas Moffett, a doctor and writer on health, thought so in 1655 in *Healths Improvement: Or, Rules Comprising and Discovering the Nature, Method, and Manner of Preparing All Sorts of Food Used in this Nation.* Here diseases are classified by their nutritive remedies and the difference between the chewable and nonchewable remarked. But Moffett anticipates and La Mettrie repeats an old saw about the diet of the ancient Israelites, claiming that their abstinence from meat was a sign of their ferocity: *"Blood* being the charet-man or coacher of life, was expressly forbidden the *Israelites,* though it were but the blood of beasts, partly because they [i e., the Israelites] were naturally given to be revengeful and cruel hearted, partly also because no blood is much nourishing out of the body, albeit in the body it is the onely matter to true nourishment."[12] So the old isomorphism about raw meat and ferocity long antedates the European Enlightenment. Here it is raw meat as the purveyor of

---

[12] Thomas Muffett, *Healths Improvement: Or, Rules Comprizing and Discovering the Nature, Method, and Manner of Preparing All Sorts of Food Used in this Nation. Corrected and Enlarged by Christopher Bennet, Doctor in Physick,* (London, 1655), 138, an important work for the history of medical consumption or phthisis.

blood that counts; if it had been cooked, the blood might have been cooked out of it. Muffett also calls "milk" a type of meat—so nutritive is it—and claims that only the most experienced physicians know how to prescribe the correct amount each patient should consume.

II: Connoisseur of Consumption and Waste: Doctor Gideon Harvey

There is no need to survey all the treatises about meat in the early period—dietary and medicinal—because Gideon Harvey summarizes their views, especially about the raw and cooked in relation to health and commerce. Harvey was probably born in Holland in the decade before the English Revolution began, born that is in the 1630s; he was medically educated in Oxford and Leiden (already the leading medical school in Europe), well travelled, and he eventually practiced in London for most of his life. The tract about meat is his *Morbus Anglicus: Or, The Anatomy of Consumptions. Containing the Nature, Causes, Subject, Progress, Change, Signs, Prognosticks, Preservatives; and several Methods of Curing all Consumptions, Coughs, and Spitting of Blood* (1666), a discussion of medical and economic consumption.[13] This is an intuitive meditation and ingenious analysis by a shrewd social and medical commentator whose career had not yet become checkered by poisoned relations to the College of Physicians in London. Written after Harvey finished his "Discourse of the Plague" in 1665, just before the Great Fire raged, Harvey teases out the analogy of meat and national stereotype early in the English Restoration.

He begins by explaining how "Physicians in their Physical discourses, make use of several names, which are all translated into

---

[13] Harvey's biography awaits compilation and analysis. He was neither saint or sinner but one of the most interesting figures of the English Restoration, always brimming with ideas, even when they were wrong. The old *DNB* entry is riddled with errors and short on interpretation, especially for failing to credit his intuition: it worked overtime and rarely abandoned him.

this one word of a *Consumption*, as if they bore no different significations; such are *Phthisis*, *Phthoe*, *Tabes*, *Morbus tabisicus*, *Marcor* [sic], *Marasmus*, a Marcid Fever, an Hectick Feaver, and an Atrophia."[14] He pleads for discrimination among the types of wasting, such as that "the said *Marcour* may likewise be caused by Famine, or over abstinence from food."[15] Nevertheless Harvey has another purpose: the demonstration that British diet differs from other countries' and the degree to which the English diet causes illness. Specifically, "We [British] differ extremely from all others in our dyet. *Flemmings* and *Germans* buy flesh meat by the pound, and eat it by ounces; we buy meat by whole joynts, and eat it by pounds."[16] His nations are located north of the Alps; the countries themselves are insignificant: the point is that other northern countries consume only a fraction of the meat of the English.

Quantity is flanked by preparation: "They [the Europeans] usually boyl and roast their meat, until it falls off from the bones; but we [English] love it half raw, with the blood trickling down from it, delicately terming it the Gravy, which in truth looks more like an ichorous [sic] or raw bloody matter."[17] This was a passage in a book— Harvey's *Morbus Anglicus*—Samuel Johnson knew well and cited as the primary example for "gravy" in his *Dictionary*.[18] "Gravy" was an old word by 1666, when Harvey wrote; a medieval word that came into its own among the Elizabethans to denote the fat and juices that exude from meat. Harvey further contributes to it making the point that the food the English think is "gravy" is actually "a delicately termed" raw type of "gravy." Raw blood or cooked gravy merge, the one indistinguishable from the other.

"Whilst in Europe," Harvey claimed, "only the wealthy [are] able to eat a lot of meat, in England 'on the other hand great and small, rabble and all, must have their Bellyes stuffed with flesh

---

[14] Harvey, *Morbus Anglicus*, 4.
[15] Harvey, *Morbus Anglicus*, 6.
[16] Harvey, *Morbus Anglicus*, 76.
[17] Harvey, *Morbus Anglicus*, 76.
[18] Samuel Johnson, *Dictionary of the English Language* (London, 1757), entry under "gravy. 1."

meat every day, and on Sundayes cramb their guts up to the top with pudding."[19] The modern institution of the Sunday special meal is, Harvey suggests, already in place and lorded over by "flesh meat" even among the poor. It won't be necessary to await Hogarth's beefeaters and clubbish Beefstakes; they are already here early in the Restoration. Harvey continues to explain how the poor obtain their meat, its costs, and—again—its rawness. One can see them sinking their one or two teeth, or perhaps their wooden teeth, into this juicy rarebit; the muddy ponds between their teeth already clogged with the discolored debris of former Sundays. Mastication and gesture have been far more pronounced than anything we know in our refined versions that seek to disguise the chewing even among the crudest among us. Indeed Harvey's description of the Sunday meal culminating in the guts crammed full of pudding is the best that survives.

Harvey's analysis contrasts the diets of England and the Continent principally to show its effects on national character and (or what we would call) stereotype. Thus, he writes, "we have parallel'd the dyets of two Nations, in order to a further examination of their different effects, rendering those [the Europeans] of a squabbish lardy habit of body [i. e., flabby, pudgy, lethargic in looks]; us [the English] of a thinner though more fleshy appearance, and some who by their stronger natures, exercise, or labour, are equally matcht to digest and subdue that mass of flesh they daily devour, acquire a double strength to what those Hermits receive from their Herbage."[20] Harvey's materialism may shock us—the notion that we *are* what we consume; but this materialism remained a crucial component of national stereotypes handed down in the thought of Montesquieu, Hume (the Hume who remembered when annotating his own "Political Essays" that "an eminent writer has remarked, that all courageous animals are also carnivorous, and that greater courage is to be expected in a people, such as the English, whose food is strong and hearty, than in the

[19] Harvey, *Morbus Anglicus*, 76.
[20] Harvey, *Morbus Anglicus*, 77.

half-starved commonality of other countries"), Buffon, Linnaeus, and many encyclopedists pronouncing on national characteristics.[21]

Hence two semi-philosophical discourses, one could say, developed in the period 1660-1760 about the eating of meat. Locke (who was a doctor as well as philosopher) and the Lockeans claimed, on the one hand, that the consumption of meat and drink (the substances) were natural rights no less than the ownership of property. Benjamin Franklin and the Americans believed, in contrast, that meat-eating was "murder of a very definite type" rather than a natural right. The philosophical argument was conducted from the view of effect as well as cause. That is, the effect of meat consumption was crucial to the natural, or unnatural, "right" it apparently was. But if the British eat more raw meat than the Europeans, they also used it up better, as Dr. Harvey claimed. Their health? "Wee'l [sic] insist a little further upon the matter," Harvey writes, "first, touching our so greedy devouring of flesh, especially Beef, and Mutton, whereof there is a greater quantity consumed in *England*, than in all *Spain*, *France*, *Holland*, *Zealand*, and *Flanders*."[22]

Harvey's explanation of the rise of disease from the consumption of nearly raw meat is detailed and thorough. "Lastly, know," he says, "that flesh meat being so nutritive, and likewise hard of digesture, doth abound with the most and worst dregs of any other kind of meat [i.e., fish, fowl, game, etc.], especially if not totally digested, as seldome it is by those that glut down such immeasurable proportions of flesh." The language is specific: the excessive "immeasurable" and the comparison with other flesh. Hence meat of inferior quality, uncooked, "glutted down" rather than adequately masticated (as if those inferior and few teeth could chew anyway), and consumed in huge quantities irrespective of social class.

---

[21] I am indebted to Dr. Caroline Warman, my Leverhulme Trust postdoctoral fellow, who is writing about this subject.
[22] Harvey, *Morbus Anglicus*, 78.

It is possible, of course, that the account is not historical; a fiction Harvey conjures as a fantasist to please one or another type of audience. But to what end if the target is *all* the social classes, rabble to rich? And why pounce on flesh meat when he was not known to have been a vegetarian or spokesmen for the fisheries or farmers of game and poultry? It is more likely, given Harvey's context and the rhetoric of persuasion in *Morbus Anglicus*, that he believed what he says of meat-eating and its medicinal consequences. For "These dregs immediately perfuse [sic; i.e., enter into the blood stream] the blood with melancholy, cause obstructions of the Spleen and Liver, and stick in the capillar insertions of the Stomach, being soon incinerated and calcined into such Salts as we permitted in the preceding Chapter: which after a short interlapse of time produce Coughs, Phthisicks, and at last a Pulmonique Consumption."[23]

Here the long historical curve of medical consumption must be borne in mind. For *Le grande chose pulmonaire*, as the great Paris doctors were to refer to pulmonary tuberculosis at the end of the eighteenth century, was historically the *end* of a process that began in meat diet. Harvey, unlike Cheyne and the herbalists of the next generation, was not promoting a vegetarian diet, but explaining how raw meat ended up in pernicious *tabes*; and—even more drastically and pressingly—drawing the analogy between meat-eating and economic acquisition, as we shall see: the first, I think, to do so unequivocally in the English tradition. "If God had thought flesh meat onely best for us, he would never have provided all these other Creatures, as fish, and herbs, for mans food (all things being created for him) unless necessary to be eaten with other victuals; for flesh and fish single would otherwise have been sufficient; besides, *God and Nature doth nothing in vain.*"[24]

Harvey's reasoning is more intuitive than logical or learned. He leans on his belief that we are what we eat: medically and economically a British nation of consumers. Moreover, the cosmic

---

[23] Harvey, *Morbus Anglicus*, 78.
[24] Harvey, *Morbus Anglicus*, 104.

order is integrated by food consumption just as a balanced diet is, and just as the body is wasted by these same medical consumptions. For this reason, the balanced diet is the best. Hence his rules for health: "Eat flesh meat four or five times a week; and fish twice or thrice', but 'Never make a meal of flesh alone, but have some other meat with it of less nutriture, as in Summer, Peas, Bears [sic, Pears?], Artichoaks [sic], Salats [sic], *etc.* in the Winter Butter'd Wheat, Milk Pottage, Broaths, or *Souppes."*[25]

Persons "wasting away" must be even more cautious, for those in consumptions of any type amounted to practically half the medical classifications of the Restoration.[26] "Consumptives" should "Abstain from all obstructive, melancholique, and dreggish Victuals; as Beef, Pork, Geese, Ducks, Cheese, Crusts of Bread, Pyecrust, pudding, Saltfish, hard boyl'd or fry'd Eggs, or any kind of fry'd Meat."[27] The "frying" of meat was almost as injurious as eating it raw, although fewer ate it fried than raw. "Feed onely upon meats of easie digesture, and inclining somewhat to a moist temperature; as Veal, Chickens, Poulets, Mutton, Lamb, Sweetbreads, Potch'd Egs [sic], *etc.* and among the sorts of Fish, Soals, Whitings, Perch, *etc.* among herbs, Lettice [sic], Endive, Succory, Sorrel, Porcelain, Chervil, *etc.* [sic] but note that they must be boyl'd."[28]

Across the Channel the French addressed these points from the vantage of cookery and refinement, by describing the civilities and health-producing balms of their cooked foods in contrast to the raw British ones. Contemporary French discussions were many, none more profusely cited then than Louis Lemery's of 1702, soon translated into every language in Europe, English in 1704, and frequently issued.[29] His work became the bible of food

---

[25] Harvey, *Morbus Anglicus*, 105.
[26] This is a difficult point to document except in light of the alternative medical conditions: gout, dropsy, fever, agues, nervous disorders. Dysentery was minor in this respect, as were cholic disorders.
[27] Harvey, *Morbus Anglicus*, 108.
[28] Harvey, *Morbus Anglicus*, 108.
[29] I use the first English translation: Louis Lemery, *A Treatise of Foods, in General* (London, 1704).

as nourishment in the first half of the eighteenth century. Lemery was, as the titlepage describes, the "Regent-Doctor of the Faculty of Physick at Paris, and of the Academy Royal of Sciences," and still practiced according to the old Galenic humoral pathology. Hence the best diets were tailored to suit the individual constitution; phlegmatic, or moist, constitutions requiring more meat to absorb the moisture than those of dry or "bilious constitutions" who were told to eat less meat.[30] Lemery's underlying agenda was the incorporation of diet into national stereotypes, showing that the French have the most "refined Cookery" and therefore also have a "fine and agreeable Taste" which permits them to surpass other nations "in every thing else," and in toto "excell all Nations" in everything.[31]

Lemery is resigned on the matter of flesh-eating—all flesh—even if the eating of meat is potentially the most pernicious. His view in brief is that civilized society has been flesh-eating for so long that it would be impossible to deflect it now. However, "if it [i. e., meat] had never been used, and that Man had been content to feed upon a certain number of Plants only, it would never have been the worse for them."[32] Thus if we had all

[30] Lemery, 139: "As our Stomachs could never endure the raw Flesh of Animals, they boil, roast, or fry it, in order to make the easier Digestion...Boiled Victuals being moister than that dress'd otherwise, agrees best with those that are of a dry and bilious Constitution, and that are inclined to be Costive. Fried and Roasted Meat on the contrary, is more suitable to those of a Phlegmatick temper, to such as abound with superfluous moisture, to those that are subject to Rheums, and Distempers of that kind."

[31] See also this accompanying passage in Lemery, 142: "The use of Animals for Food varies according to the People and Country: In short, this we are sure of, that we have some here that can never eat of them but in pressing Necessity, because we have got a strong Aversion for them, tho' we know not why. In the mean time these same Animals are earnestly sought after in several Places; Again there are several others which we make no use of here, because not to be had; they [143.] eat them with Delight in those places where they breed....The Poles, Germans, and English, whose Countries afford good Pasture, breed all sorts of Herds of Cattle; however, they value Beef and Swines Flesh before any other....The French, whose Country abounds in Necessaries, also make use of several sorts of Animals, which they prepare and dress in so delicate a manner, and with so fine and agreeable a Taste, that it may be said, they have refined Cookery, and do therein, as they fancy they do in every thing else, excell all Nations."

[32] Lemery, 146. The entire passage is well worth citing: "There are some who take upon them to shew, that the Foods we have from Animals are hurtful, and prejudicial to our

become herbalists or vegetarians in the primitive ages of history, we might be healthier than we are now (and presumably chew more easily and digest our intake too). But the moment has passed, it is too late, so we might as well eat integrated diets based on balance and moderation.

Few British *savanti* agreed. John Woodward, a much published physician in the England of Pope and Queen Anne who was an iatromechanist schooled in Newtonianism, was less concerned about meat per se than "the new gluttony,"[33] which was itself a sign of the lewd times and would lead to further immorality and, in the countryside, even lead to atheism.[34] The logic here was a

Health; that they were not made by the God of Nature for the use we put them to...They add...that, the Flesh of Animals do, by the excessive Fermentations they cause in our Bodies, corrupt our Humours, and occasion divers Diseases: They further observe, that those who feed upon gross Flesh to excess, such as is that of most quadrupeds, they become gross, stupid, and as a Man may say, acquire a resemblance of Temper and Inclination with those Animals whose Flesh they feed upon....[147.] As for my self, I am of Opinion, without entering into all these Discussions which I think to be of little use, it may be said, That the use of Animal Flesh may be convenient, provided it be in Moderation, in as much as this affords good Nourishment; however, it may be, if it had never been used, and that Man had been content to feed upon a certain number of Plants only, it would never have been the worse for them: But it's no longer a question to be disputed, and if it be an abuse, it has so long obtained by Custom in the World, that it has become necessary."

[33] John Woodward, *The State of Physick: And of Diseases; With an Inquiry into the Causes of the Late Increase of Them: But More Particularly of the Smallpox. With Some Considerations Upon the New Practice of Purging in that Disease. To the Whole Premised, An Idea of the Nature and Mechanism of Man: Of the Disorders to which it is Obnoxious: and of the Method of Rectifying Them* (London, 1718), 201: "Gluttony is sure to bring on Load of the Stomach, and Indigestion: and as this, become putrid, and vitious, is, with a redundant or degenerous Bile, the Principle and Cause of all Diseases, a large and plentyfull Crop of them must needs follow."

[34] The stomach was the focus of Woodward's attention in disease, especially in *The State of Physick*, a view that fed into the developed notion of disease primarily located there and then sent out to other territories, the lungs, bowels, brain, etc. Hence, "1. The first Scenes, and the Beginnings of all Things, good or bad, to the Body, are in the Stomach. As Impressions are made there, and Things transacted, rightly, or wrongly, the Body, baring exterior Accidents, is well, or ill: [2.] and the Health firm, or interrupted. While the Stomach prepares and dispenses down into the Blood an Aliment, that is wholey good and right, every Thing must be so over all the Habit: and the Organs supplyd, and inabled to do their Duty rightly, in Passion, Exercise, Labour, and all other Exigences of Life. But, when the Fountain is puddled, the Streams must be so too: and Matter, vitious, and erroneous, in the Stomach, must unavoidably be diffused over and incommode the whole Frame. So that our first and chief Aims, either for the Preserving of Health, or for the Restoring of it, ought to be directed to the Stomach."

sequence leading from gross material consumption of a product to luxury, effeminacy, and debauchery. The cities were already in decline, mired in debilitating illness, for "This is certain that London has been much less healthy, of late, than heretofore: and Diseases generally more mortal."[35] In Woodward's view all excess was to be avoided, especially in diet. The "New Cookery," he believed, imported from abroad (France) was so heavily spiced and dripping in sauces that bodily digestion was impaired. Nothing could ruin countries more certainly than diet. In sink with the xenophobia of other physicians of the reign, Woodward thought food the sure route to disease and early death. Woodward used his mechanics to explain the physiology of digestion. He adumbrated how the scholar's diet caused him to be lank and of "consumptive habit;" an idea long developing since the Renaissance before it climaxed in the Romantic period in such sickly figures as John Keats and the poet-as-consumptive cult.[36] Above all was the ruinous way that diet controlled the passions,[37] meat most of all, and that foreign foods were changing the national character of the English:

> to the Neglect of the much better and more wholesome
> Products of our own Country, the Mispending our Trea-
> sure, and carrying it even to the most distant and remote
> Parts of the World: or to the Exchange of our own usefull
> Manufactures, not only for Trifles, and Things of no real
> Use, but such as are detrimental, and injurious. To these
> Sauces, and these Liquors, our vertuous, wise, stout,
> healthy Ancestors were Strangers. By the former, Intem-

---

[35] Woodward, *The State of Physick*, 192.

[36] Woodward, *The State of Physick*, 5: "For which Reason Scholars, and Men that give themselves up much to Thought and Study, are ordinarily lean and consumptive."

[37] Woodward, *The State of Physick*, 200: "What will be likely to push on, and hasten this, is the Exorbitance of Passion, which must necessarily attend the Increase of the Biliose Salts, that are the Instruments of the Passions, drawn out of the Meats eaten: and must be consequently proportion'd to the Excess of them. Then Meats seasoned, Sauces, and Things of high Savour, which make so considerable a Figure in the New Cookery, supply those Salts in the greatest Plenty."

perance and Excess, is promoted: and consequently, in Tract of Time, an Indigestion brought on.[38]

Food and the passions were intrinsically connected in this material way; more accurately stated, the passions and the act of "consuming" specific foods for specific purposes. Food and passion existed in direct relation to each other. Thus the passions rise, Woodward claimed, to derail human beings and corrupt them, and:

> What will be likely to push on, and hasten this, is the Exorbitance of Passion, which must necessarily attend the Increase of the Biliose Salts, that are the Instruments of the Passions, drawn out of the Meats eaten: and must be consequently proportion'd to the Excess of them. Then Meats seasoned, Sauces, and Things of high Savour, which make so considerable a Figure in the New Cookery, supply those Salts in the greatest Plenty.[39]

Woodward's ideas of 1718 did not change significantly in the next decade when Swift (who could have read Harvey's *Morbus Anglicus*) was using meat for yet other purposes in *Gulliver's Travels* and *A Modest Proposal*, as his basis for comparison among the Houhymhymns and Yahoos, and among Irish parents eager to sell their juicy red babies for a profit. Survey the decade of the 1720s and you find prohibitions everywhere against raw "sharp meats."[40] The notion that "sharp meat," which included the "raw," could prove to be the site of origin of fatal, even wasting-away, diseases is alien to us, but was not to Defoe's and Swift's world. Certainly

---

[38] Woodward, *The State of Physick*, 194.

[39] Woodward, *The State of Physick*, 200.

[40] For example, an anonymous work entitled *The Best and Easiest Method of Preserving Uninterrupted Health to Extreme Old Age: Established upon the Justest Laws of the Animal Oeconomy, And Confirmed by the Suffrages of the Most Celebrated Practitioners Among the Antients and Moderns. From a Manuscript Found in the Library of an Eminent Physician Lately Deceased, and by him intended as Legacy to the World*(London, 1726, 2nd ed.) claims that "this Distemper may proceed from a total Suppression of Evacuations, or be induced by any great Force upon the Lungs from accidental Causes, or by hot and sharp Meat" (116).

not the Swift for whom the whole of Ireland was but a grazing pasture for the English: an idea he bitterly developed in satiric tones in *Gulliver's Travels*.

Just so seemed these matters to jolly, Falstaffian, Dr. George Cheyne, who could be as depressed as he was merry. His *English Malady* of 1733 is now too well known to require an epitome except to say that it was the compendium of a decade's writing on food and diet in relation to health and temperance.[41] Cheyne was not a vegetarian at all but a temperist and neo-Stoic whose rock of belief was abstinence, balance, and moderation. He promoted with religious fervor and in sweeping Miltonic rhetoric a "lettuce and seed" diet as the antidote to the murderous gluttony of juicy (and perhaps raw, if the French charge had any validity) beef diets:

> When I behold, with Pity, Compassion, and Sorrow, such *Scenes* of Misery and Woe, and see them happen only to the *Rich*, the *Lazy*, the *Luxurious*, and the *Inactive*, whose who dare daintily and live voluptuously, those who are furnished with the rarest Delicacies, the richest Foods, and the most generous Wines, such as can provoke the Appetites, Senses and Passions in the most exquisite and voluptuous Manner: to those who leave no Desire or Degree of Appetite unsatisfied, and not to the *Poor*, the *Low*, the *meaner Sort*, those destitutes of the Necessaries, Conveniencies, and Pleasures of Life, to the Frugal, Industrious, the [29.] Temperate, the Laborious, and the Active: to those inhabiting barren, and uncultivated Countries, Desarts, Forests, under the *Poles* or the *Line*, or to those who are rude and destitute of the Arts of Ingenuity and Invention. I *must*, if I am not resolved to resist the strongest Conviction, conclude, that it must be something received into the Body, that can produce such terrible Appearances in it, some flagrant and notable,

[41] George Cheyne, *The English Malady: Or, A Treatise of Nervous Diseases of all Kinds, As Spleen, Vapours, Lowness of Spirits, Hypochondriacal, and Hysterical Distempers, &c.* (London, 1733).

Difference in the *Food*, that so sensibly distinguishes them from these latter. And that it is the miserable Man himself that creates his Miseries, and begets his Torture, or, at least, those from whom he has derived his bodily *Organs* [i.e. his parents].'[42]

Cheyne's view exceeded the proverbial one in the health diagnosis of his time (and the philosophies of materialism abetted it) that we *are* what we eat. He thought we abrogate our free will when we indulge, and must be held responsible for the misery we suffer. He judged British history, in part, through the lenses of diet: "*Milk* and *Honey* was the Complexion of the *Land of Promise*, and *Vegetables* the *Diet* of the *Paradisiacal State*: And...such a *Diet* will (if any thing) certainly cure, by the Concession of all *Physicians*, learned and unlearned, ancient or modern, *High* or *Low-livers*, the *Gout*, the *Consumption*, and the *Scurvy*, and such like atrocious, otherwise incurable [sic] and mortal Distempers."[43] For Cheyne the "English Malady" was all that constellation of melancholy, even "self murder," predicated upon the constants of English climate (the climate forever feeding into national stereotypes), the six non-naturals (air, water, earth, sleep, motion, diet), and the national temperament these produced; but if climate could not be changed, diet and sleep could, and Cheyne believed that diet determined the quality of human sleep, not the reverse.

Digestion or its antithesis, indigestion, was another key concept in these discussions. Poor digestion caused constipation, sleeplessness, and nightmares; no digestion produced illness confirmed by the mechanistic mathematics these post-Newtonian doctors produced. Evaluations of the raw and cooked therefore were made not merely as cultural smoke screens amidst a dialogue aiming to define the nation Albion in its ascendancy over France, but as the surest routes to health. In England the Augustan "digestion doctors" demonstrated (building on Sanctorius and the

---

[42] Cheyne, *The English Malady*, 28-29.
[43] Cheyne, *The English Malady*, 368.

temperance doctors of the seventeenth century) that too much matter in the colon led to fermentation of the bile and self-poisoning through constipation. Raw meat was especially culpable here because tachyphagia, or the hasty eating of carnivores, left so much undigested food in the system that poisons from the putrefying meat formed in the intestinal tract. For this reason specifically, the doctors of the 1760s advocated the taking of red wine with meat (whether raw or cooked), as others did for the advocacy of conversation, on grounds it would slow down the onset of tachyphagia.[44] Hogarthian gluttons who treated the body as a dump would pay the price. Graciousness and gentility did not go hand in hand.

Hence it was only a short route from the body as dump to the digestive canal as common sewer. Here, accumulating in the gut, was the body's veritable putrefactive apparatus suffused with fermenting alkaloids, bile and chyle. Even a small amount of putrified meat in the distended stomach was sufficient to kill, as M'lady this and M'lord that learned to their peril after a night out on the town. No wonder then that the Georgian gut—for the first time in medical history[45]—became the supreme seat of hypochondria: not merely because the gut was anatomically positioned beneath the hypochondrium in the pulmonary cavity, but because its routine distension—the repetitive pulling of the stomach out of shape over a long time—could lead to perilous maladies. Plot the full trajectory of Enlightenment hypochondria, from Cheyne (who suffered an idiopathic variety) to Samuel Taylor Coleridge (perhaps the most pronounced literary hypochondriac before the advent of Proust) and you see how often this fear of the gut reduces to the terror of meat putrifying in the stomach.

[44] For wine, see the writings of Dr. Edward Barry (1696-1776), who had written a medical thesis on food's nourishment before launching out in the history of wines; for conversation see Boswell's account of the dining club of Samuel Johnson in the *Life of Johnson*.
[45] See G. S. Rousseau and David Haycock, "Samuel Taylor Coleridge: Hypochondriac?" (forthcoming), which deconstructs Coleridge's obsession with his own gut and constructs the early modern background from c. 1750 forward to the mid-nineteenth century.

Dr. John Arbuthnot concurred, not merely as the addressee of Pope's famous satiric poem about aberrant pests like "Sporus, that Curd of Asses' milk," but also as the author of *The History of John Bull* whose "honest, plain-dealing fellow" is "choleric and bold," in large part as the result of his meat diet. Arbuthnot also wrote a much-read treatise on food—*An Essay Concerning the Nature of Aliments* (1735), a study of the forms of nourishment.[46] His was the second work on food Samuel Johnson cited as endowed with authority in his *Dictionary*, the first being Gideon Harvey's *Morbus Anglicus*, as we have seen. Arbuthnot listed the different types of cookery of flesh according to states of digestion and made them the basis for alimentary health.[47] If we continued in this chronological mode down through the 1750s and 60s, the validity of the paradigm about the raw and cooked in relation to digestion would continue as an *arriére pensée*, as does the idea that the English like their meat rawer —redder—than the Europeans.

III: Enter the French to Challenge the British

The French did not cease to ridicule the British after Lemery, either for the superiority of their French cookery or the barbaric meat-eating customs of the Brits. The line from Lemery to Voltaire and La Mettrie occupied a half century, but as the epoch evolved, the symbolic role of cookery changed, and by the revolutionary fin-de-siècle food signified in ways it had not a century earlier. French scholars Flandrin and Montanari, who have studied the dietary

---

[46] *An Essay Concerning the Nature of Aliments, And the Choice of Them, According to the Diferent Constitutions of Human Bodies. In Which the Different Effects, Advantages, and Disadvantages of Animal and Vegetable Diet, are Explain'd*, 3rd ed. (London: 1735).

[47] Arbuthnot, *Aliments*, 108: "Preparations by Cookery of Fish or Flesh ought to be made with regard to rectifying their most noxious and slimy Substances, and to retain those that are most nutritious; such Preparations as retain the Oil or Fat, are most heavy to the Stomach, which makes bak'd Meat hard of Digestion. Boil'd Flesh is more moistening, and easier of Digestion than roasted."

patterns of the *ancien regime*, illuminate the matter in their acclaimed 1996 history of food.[48] Cookery changed from a utilitarian function in the provision of nutriment and the sustenance of health to an art of pleasure calibrated to the mood of the occasion. But food also assumed symbolic meaning, especially within the sign system extending from individual character to national stereotype in nation-states and was an intrinsic part of that national stereotype. The road from "we are what we eat" had been extended: it was now that you are a Frenchman (for example) because of the *way* you eat. Sharper yet: *what* you eat, and the *way* you eat it, reveals whether or not you *can* be French.

For instance, in Prévost and Crebillon *fils* food is barely mentioned. By the time of *La religieuse*, written in 1760 but not yet published, things have changed, and while there is no meat, food has been elevated to the new significance. Rousseau was the first, however, to provide detailed explanations of the "characters" of foods, almost as if they were nations; gendering them (meat is male, dairy and sugar female) and sorting them by national characteristics. Remarkably, Laclos' *Les liaisons dangereuses* (1782) is perfunctory about food and omits food as part of seduction rituals, but it is Sade, predictably, who provides the most precise picture of the consumption of meat, raw and cooked: a canvas at once perverse, pleasure-related, and brutal.

Flandrin and Montanari have said, in a sense, the last word when comparing French and English food rituals in the eighteenth century. They show how the French consume poultry and boil and overcook their meat; the English, in contrast, are meat eaters and consume it almost raw—the old charge. So rare that by the time Horace Walpole traveled on his Grand Tour in the 1730s, and Tobias Smollett a generation later in the 1760s, neither claims to be able to find any good meat in France.[49] But even Flandrin and Montanari are less nuanced in their presentations than the great French writers themselves. In *Manon Lescaut* supper forms the

[48] Jean-Louis Flandrin and Massimo Montanari, *Histoire de l'alimentation* (Paris: Fayard, 1996).
[49] Flandrin and Montanari, *Histoire de l'alimentation*, 676-77.

prelude to betrayal but it is a menuless meal. Diderot's lesbian nun, however, lures calculatingly with "fruit, marzipan and preserves."[50] The nun's consumption of edible luxuries provides the best clue to her frivolous and pleasure-loving character. But even the nun pales in comparison to Julie's ex-lover Saint-Preux in Rousseau's *New Heloise* who genderizes meat and calibrates sexuality to its terms. Saint-Preux, now the tutor of Julie's children, proclaims didactically:

> Indeed, I have observed in France, where women live all the time in the company of men, they have completely lost the taste for dairy products, the men largely that for wine, and in England where the two sexes are less confounded, their specific tastes have survived better. In general, I think one could often find some index of people's character in the choice of foods they prefer. The Italians who live largely on greenery are effeminate and flaccid. You Englishmen, great meat-eaters, have something harsh that [373] smacks of barbarity in your inflexible virtues. The Swiss, naturally cold, peaceful, and simple, but violent and extreme in anger, like both kinds of food, and drink both milk and wine. The Frenchman, flexible and changeable, consumes all foods and adapts to all characters. Julie herself could serve as my example: for although she is sensual and likes to eat, she likes neither meat, nor stews ...[51]

Rousseau even proclaims that in Britain "great villains harden themselves to murder by drinking blood."[52] Here then, again, is the old Voltairean mythology about raw meat come home to haunt the English: an uncouth, barbaric, almost genderless, male people (this point intensified in other guises in the Victorian era),

---

[50] Denis Diderot, *The Nun (La religieuse)*, tr. L. Tancock (Harmondsworth: Penguin, 1977), 123; the work was written but not published in 1760.
[51] Quoted in Flandrin and Montanari, *Histoire de l'alimentation*, 372.
[52] *Emile*, Book 2, (Paris: La Pléiade), IV: 411.

according to the softer Continentals. Yet, in the interstice between Rousseau and Sade is found the genuine coin of this carnage. In the *Hundred and twenty days of Sodom*, La Duclos, the first prostitute-storyteller, recounts how her client D'Aucourt altered her diet to poultry as part of a regime to satisfy his coprophagia. He indulged his white-meat fetish by restricting her diet to it: a rich man's craze for luxury, connoisseurship and perversity interlinked by this meat.[53] In *Justine*, the blood-sucking libertine Gernande bleeds his wife while raping her. After she dies, Gernande and his nephew Bressac place her body in the centre of the dining table as if ready to devour her still fresh meat.[54] Beatrice Bomel-Rainelli, writing in an illuminating book of essays edited by Beatrice Fink, the distinguished Sade scholar, claims that the correct way to construe this libertine banquet is to view Gernande and Bressac as cannibals sucking the newly killed animal.[55] In the *History of Juliet*, written in 1801 but hardly to greet the new century, it becomes clear that meat—all meat, including human meat—is fair game for consumption. The Russian Minski—Juliette's host—and Italian Sbrigani are libertines; men of unbridled savagery, almost cannibalistic: wild, unrestrained, larger than life.[56] They enumerate all they devour, with the relish of a fetish. Meat-eating represents their grossest animal appetite: as blood is drunk, so human flesh is preternaturally consumed. Each act of mastication coexists within cycles of pain and torture (including sexual orgasm) which enhance the rituals of pleasure.

Casanova was as fetishistic as Sade but not in the consumption of meat. Venetian to the end, he preferred fish and oysters. His pallate was formed by the diversity of excess and creased in the strongest taste butts. He particularly devoured foods with strong tastes: deer when it had what we call 'haut gout,' cheese

---

[53] Sade, *Les cent-vingt journées de Sodome*, ed. Michel Delon (Paris: Gallimard Pleiade, 1990), 180.

[54] Sade, *La Nouvelle Justine*, ed. Michel Delon (Paris: Gallimard Pleiade, 1995), 862-870.

[55] Beatrice Bomel-Rainelli, "Sade ou l'alimentation generale," *Dix-huitieme siecle*, 15 (1983), 205.

[56] Sade, *L'Histoire de Juliette*, ed. Michel Delon (Paris: Gallimard Pleiade, 1998).

when ripe almost to the point of putrefaction, especially if his naked eye could watch the little vermin dancing around. He relished ragout, fricassée, birds (grouse, larks, doves), hare, roastbeef, grilled filet steaks, Turkish pilaw (rice with male lamb), wild rabbits, meat soup ("bouillon" or "consommé") when cooked to excess and covered with animaculae and bugs. In England he complained bitterly about the lack of soup—this because the English ate their meat raw and did not waste it, as did the French, in soups. He liked to eat cold cooked beef which was made popular by Louis XV, despite the criticism of then conservative French gourmets who ate beef only when fried over a fire on a spit. A Venetian in his taste butts, he preferred fish to meat; and found himself lured to sexual indulgence by oysters and visually spectacular meal arrangements containing many courses, the diversity of the meal symmetrically anticipating the diverse sexual acts. This symbolism of excess, opulence, the cooked in preference to the raw, cannot be missed in his culinary presence.[57]

Casanova's mastication should not be overlooked. He lost his teeth around sixty, replacing them with an expensive porcelain set. The new false pair appears not to have diminished the vigour of his pallet, or mastication, and afforded him a new excuse to relate his often-told story about his flight from the Venetian state prison, for *sans* his natural teeth he could no longer recount it with adequate pronunciation. The main point, then, to draw about the raw and the cooked from the hundreds of pages he has left pertains to the necessity for his meat to be prepared in the Continental style. At the Gros Caillou, near the Invalides in Paris, he consumed his *boeuf à la mode* garnished with stripes of lard treated with basil, shallots, cloves and parsley, the whole marinated in red wine. Even when in London in 1764, where he failed to succeed in his suicide attempt, he insisted that his "roast beef à l'anglaise" should be a big piece of beef with ribs and filet roasted on a spit for 3 to 4 hours. These and many other meals in England

---

[57] I am grateful to Helmut Watzlawick of Geneva, Switzerland, for directing me to these passages in Casanova's writings.

were hardly made of the proverbial raw red meat of the consumptive, melancholic English. He had certainly tasted the cold cooked beef popular among travellers in France and Switzerland but his taste was for roasted meat which the French—perhaps spiraling downward from their sovereign—greatly preferred.

French literature at large reflects the century's massive swing from the crude and the raw to the roasted and the cooked, increasingly embodying symbolic descriptions of the preparation of meat as the century wore on. The mythology about the English may, therefore, have been grounded in some reality. If there is less amplification in Prévost and Crebillon *fils*, by the time Diderot published *La religieuse* in 1760 and Rousseau his *New Heloise*, meat has taken a more central place. Rousseau could not dislodge the English from their brute meat-eating, perhaps as the result of memories of his own sojourn there. *Emile* indicts the British for their food more cosmically, proclaiming (as we saw) that in England "great villains harden themselves through blood."[58]

IV: Conclusions

Meat—the product, its preparation, consumption, symbolism and mythologies—was merely one of the Enlightenment's consuming passions yet was crucial to the formation of its national stereotypes. Its production and consumption was inherent in the construction of national identities, national diseases, political economies (Swift had revealed to the world most of what it needed to know about the politics of meat), and—not least although somewhat bathetic by virtue of its comic associations—the state of the teeth in different countries: topics then of the first importance.

---

[58] In 1704, when Locke died, cooking in England fulfilled a more or less simple function; by 1771, when the *Dictionnaire de Trevoux* appeared in Paris, popular perception of the role of cooking had changed rather completely from a way of preparing food for eating to satisfy human needs, to an art and concommitant pleasure.

The representation of meat was as revelatory: not merely the visual tradition in Hogarth's "Beefsteaks" (the engraving) and the genre of eighteenth-century painting that deals with the hunt and its scent for meat, but as prominently in the literature of the era, as we have seen. Meat was intrinsic to the diet and health of nations far more so than the bread and sugar of the Renaissance, even moreso than it is today despite our crises of BSE and CJD because the consumption of meat has proportionally (per capita but not absolutely) diminished so much in the Western world. The Enlightenment's ranking diagnoses of pulmonary consumption, dropsy, fever, gout, all pronounced on the pernicious role the excessive consumption of meat had played, especially when served raw (if it was served raw) in the common habit of the era.

Nor can the Enlightenment body and its different sexualities be considered apart, as we have seen, from this most fundamental of material substances. Bodies primitive and civilized, fair-skinned and dark-skinned, male and female, whose material textures—dermis and interiors—glowed according to the diet and exercise they had received. In our generation we have heard much about the Enlightenment body and its sexualities, but much less about the foods—especially the exotic foods—those bodies and their representative sexualities processed. We know, for example, almost everything the members of Johnson's dining club said when they gathered but much less about the foods they ate. We know how the heroines of our novels starved themselves (Clarissa virtually unto death), but much less how they ingested their meat to produce a commensurate sexual response: one river of raw-blood fluids exchanged for another. We have already combed more Enlightenment medical treatises than any other scholarly generation in history, but we still harbour rather an unclear sense of how the physical bodies of our subjects—biographical and fictional—wasted away in one type of illness or another, when their own "meat" became diseased.

Finally, our contexts for these consuming passions and the barometers of mythology and reality. We may aim to cultivate reason above sentiment in our current lives or in our approach to our academic field, but I doubt that any among us still believes, or teaches students, that the Enlightenment was an Age of Reason. We know it was equally an Age of Passion riveted by passions of every type, which were both greater in extent and more excessive in degree than any reason; and I have suggested that one consuming passion—the consumption of meat—was then transformative for the way it combined several realms, not least the materialistic one that impinged on national stereotypes. Few topics other than raw and cooked meat can demonstrate how its men and women actually constructed the national identities and stereotypes on which they were basing their daily lives. To the French, the English habit of routinely eating raw meat revealed everything *significant* about the British: their ferocity, their uncouthness, their dead palate, their cruelty to their children, their weather (which may have been given as one further reason why they ate more meat and perhaps ate it raw: to keep themselves warm), their innate melancholy, their propensity to self-murder (suicide), their diseases. Gout may have been a pan-European affliction in the eighteenth-century, as Roy Porter and I showed in our book about the gout diagnosis, but it was nonetheless primarily an English condition: began in England, ended there, with more cases there than anywhere else, and every gout doctor of the Age of Pope or the Age of Keats knew that one of the first questions he had to ask his patients dealt with the consumption of meat.

At stake then in Enlightenment meat-eating and its masticatory pleasures is the semiotic component as well as the materialistic: the consumption of fine cuts of meat were no less "materialistic" than other types of collectables, this in the first era to bring economic consumption and philosophical materialism so proximate. If the eating of meat was not of interest entirely for itself (in the sense that the Caribbean sugar trade was then of paramount concern for economic historians, or the tea trade of the Orient and the spice trade of the East India routes), then meat-eating functions as one of the more prominent anthropological signposts of the era: certainly as indicative as clothing, housing (the country house), gardening, transportation (carriages and coaches) or education (school houses and school teachers). Meat and mastication are rather more primitive enterprises despite meat's civilizing tendencies (in the sense of Norbert Elias), as we saw among the French where no sexual seduction scene could be devoid of it, it seems, by the end of the century; and among the Italians of the next century where in their operas—from *Figaro* and *Giovanni* to *Bohème* and *Tosca*—*they* shrewdly included it in their librettos because their audiences expected it.

Meat-eating, more so than the consumption of other foods, says more about the civilizing thrust than its protein benefit; is a key to the cannibalistic urge measured by its most refined thermometer. As is the tooth's bite, so too the pelvic thrust. Show me how you consume a steak—so goes the sign system—and I'll tell you how amorously poised you are. No wonder that Freud wanted to know, above all, what the great creators, the great writers, ate; as if he could plot their creative genius by ascertaining what they ingested and how they digested their foods. And Rabelais, also a doctor, was no less prurient about the meat-eating habits of his Gargantuan populaces, as Russian theorist Mikhail Bakhtin has noticed in his study of Rabelais and his world.[59] Was this the reason, one wonders, why Edmund Burke, at the height of the English hysteria over the fear that revolution in France would spread to civil chaos in England, couched his rhetoric of terror in the sustained cannibalistic metaphor of the hyena crowd? No other animal, not even the lion or tiger, has such a powerful masticatory bite as the result of its all-powerful jaw. As the Enlightenment encyclopedias stated, the hyena's grip was unique. Likewise at the earlier end—the start of the century—when another Irishman, Swift, wrote about the grip of the tooth into baby flesh. The horror of Swift's "modest message" was much less the empty piety that "eating children is wrong" (in our society we have been routinely eating them through every type of abuse) than the appalling, if forever unseen and imagined, bite of the fierce English tooth into the bloody babe's raw flesh. This maculary bite

[59] See Mikhail Bakhtin, *Rabelais and his World* (Cambridge, Massachusetts: MIT Press, 1968).

encompassed aspects of the ghastly La Mettriean gesture—the symbolic act underlying the allegedly "fierce English." English hounds showed it. English hunts exuded it. English meat-eating proved it.

The gradual ebbing and taming of this ferocity in the aftermath of the French Revolution entails another story that required—sequentially—a Romantic movement, Regency softness, and at least a tinge of decadence. Hence by the time high Victorian values entrenched themselves in the world of Darwin and Dickens, English fierceness could relent, at least mythologically construed. But by then—the mid-nineteenth century—the Americans had come on the scene across the ocean and English propaganda for its own (i. e., English) indomitability over the Americans was another matter. The new nation, the Americans, ate more meat than any European nation and ate it faster. This new phase of meat-eating and its masticatory pleasures is a nineteenth-century American story of gluttony that replaced the European narrative and cannot be told here.

PART III
THE SCRIPTS OF SCIENCE

CRACKS IN THE CEMENT
OF THE
UNIVERSE
Hume, Science, and Skepticism

Peter S. Fosl
Transylvania University

Our imagination has a great authority over our ideas; and there are no ideas that are different from each other, which it cannot separate, and join, and compose into all the varieties of fiction. But notwithstanding the empire of the imagination, there is a secret tie or union among particular ideas, which causes the mind to conjoin them more frequently together, and makes the one, upon its appearance, introduce the other....'Twill be easy to conceive of what vast consequence these principles must be in the science of human nature, if we consider, that so far as regards the mind, these are the only links that bind the parts of the universe together, or connect us with any person or object exterior to ourselves. For as it is by means of thought only that any thing operates upon our passions, and as these are the only ties of our thoughts, they are really to us the cement of the

universe, and all the operations of the mind must, in
a great measure, depend on them.
    — David Hume, "Abstract" to *A Treatise*
    of Human Nature (1740)[1]

The "empire of the imagination" is mighty, indeed. It can join and structure ideas into ever-changing combinations seemingly without end, giving rise to unicorns, extraterrestrials, nations, abstract ideas, campuses and religions. Imagination, however, can also be a great destroyer, the force that not only combines but also disintegrates, that tears vast theories and systems and worlds of ideas asunder, rending them from top to bottom. Hume himself has seemed to many an agent of this sort of destruction, wielding with merciless power the great blade of skeptical imagination, for it is upon the shoals of Hume's skeptical philosophy that the rationalistic aspirations of Samuel Clarke (1675–1729), René Descartes (1596–1650), Nicolas Malebranche (1638–1715), and the natural theologians to deduce the nature of the self, the world, morals, and God crashed.

But as this passage with which Hume closes the abstract to his *A Treatise of Human Nature* (1739–40) indicates, there is something else here—some "secret union," almost a conspiracy, some countervailing force organizing around the banner of "human nature" that militates against the otherwise unchecked power of imagination. Hume calls these counter-revolutionaries the principles of the "association of ideas." As a theorist of these natural principles and of human nature Hume presents us with another visage. Rather than destroyer this Hume appears as a naturalistic, constructive thinker who expels skepticism with the cannon of naturalism and who helps to lay the groundwork for

---

[1] David Hume, *A Treatise of Human Nature*, ed. L. A. Selby-Bigge and P. H. Nidditch, 2nd ed. (Oxford: The Clarendon Press, 1978). Book I & II were published in 1739, Book III in 1740. The "Advertisement" was prefixed to Book III of the *Treatise* (London: Longman's, 1740). Hereafter the *Treatise* is cited intratextually as "*T.*"

modern natural science.   It is the tension between these two Humes that I wish here to explore.

The tension, of course, is deeper and more complicated than the merely historical.   Indeed, great builders must often also be great destroyers.   The tension with which I am concerned here is a philosophical tension—a tension between contingency and necessity, between fragility and stability, between vulnerability and security, in some ways between being and becoming.   The thesis I will argue is that for Hume skepticism runs deeper than naturalism—that, in fact, Hume's naturalism cannot be properly comprehended unless one understands how it is alloyed with his skepticism.   Hume is, in short, a naturalistic skeptic, not a skeptical naturalist.   To the extent that the imagination per se underwrites Hume's skepticism while the principles of the association of ideas present his naturalism, it is the imagination that runs deeper than association.   And because the principles of association, particularly that of causation, underwrite for Hume natural science, skeptical philosophy runs deeper than natural science.   The principles of association may be the "cement of the universe," but Hume is well aware that this cement remains in the face of skepticism vulnerable to cracking.   Indeed, it appears to be always already shot through with fault lines neither rationality, nor custom, nor metaphysics can seal.

In order to make my case I wish to show that for Hume what is conceived as "nature" and "natural belief" is not, in the jargon of some contemporary philosophers, properly basic.   In other words, Hume's characterization of particular beliefs and conceptions and theoretical posits (like the principles of association) as natural is itself contingent although it accounts for features of the human world that appear to us stable, orderly, and necessary.   Hume always understands that what appears this way may not be, and hence he understands that what he advances as necessary and what we perceive to be so may actually be otherwise.   The fragile, contingent character of what appears as nature and nature's necessities render Humean science to be of a rather peculiar sort.   It is a science not of deductive and empirical certainties nor a form of realism that claims to disclose the essential nature of the independent world—it is, rather, a science of finitude as well as of salubrious satisfactions.

I: Contingency Rampant: Deleuze's Hume

Among those who have attended to the importance of contingency in Hume's philosophy is Gilles Deleuze. In 1953, Deleuze published his first book, *Empiricisme et sujectivité sur la nature humaine selon Hume*, the initial excursus in his series of attempts to, in his own words, sodomize the history of philosophy.[2] In Hume, Deleuze found inspiration for what he would later call his own "empiricism." What Deleuze found inspirational was Hume's insistence upon the radical contingency characterizing all relations of ideas, both ordinary and theoretical—the same contingency Hume points to in the quotation preceding this essay with his metaphor of imagination's empire.[3]

Contrary to rationalistic thinkers like Descartes and Clarke, ideas for Hume bear no intrinsic relations with one another. Structure and order must come to them from the outside, as it were. If order is not intrinsic to ideas, then disorder—or at least de-construction—must always and already remain a pervasive possibility. In Hume's view, therefore, we are steeped in contingency. Every theory, every thought, every belief, indeed every self is—at least apparently—liable to be shattered, broken down and reordered in a new configuration. Since the "science of man," as Hume calls it (*T* xv), subsumes all other sciences, contingency, for

---

[2] See my review essay of Deleuze's work: "Empiricism, Difference, and Common lLfe," *Man and World* 26 (1993), 319–28. Gilles Deleuze, *Empiricisme et subjectivité sur la nature humaine selon Hume* (Paris: Presses Universitaires de France, 1953); tr. by Constantin V. Boundas as *Empiricism and Subjectivity* (New York: Columbia University Press, 1991). See also, "Hume" (1972) in Deleuze's *Pure immanence: essays on a life*, trans. Anne Boyman (New York: Zone Books, 2001).

[3] For additional ways in which Hume's work may be thought of as postmodern consider Zuzana Parusnikova, "Against the Spirit of Foundations: Postmodernism and David Hume" *Hume Studies* 19 (April, 1993): 1–17.

Hume, pervades the entire theoretical world, including natural science. The empiricist Hume recognized by Deleuze is a stalwart citizen of the imagination's empire of contingency.

Of course, although he emphasizes the Hume of imagination as a precursor to poststructuralist thought, Deleuze is aware that Hume's work bears the marks of "modern" philosophy, as well. While relations are external to ideas, structure does come to them. Hume develops a naturalistic, associationist theory of mental activity that points to principles of a common human nature and to general and uniform rules of association which bring order to human minds. But, according to Deleuze, the "secret tie or union" rooted in these principles of human nature is, for Hume, in itself insufficient to accomplish an ordering of the human world of experience. Human nature and the associations it determines are complemented by additional structuring forces, namely those of the social order in which any individual may find him or her self situated. In other words, each particular subject is structured not only according to the individual's natural imperatives to integrate itself but also by the imperatives of the given social order to integrate that self into the variegated network composing a public world peopled by other selves.

Herein lies another layer of contingency, for in Deleuze's view social orders, unlike the principles of human nature, are themselves contingent. These factors are contingent to the extent that they are various and inconsistent, changing across space and time. Hence in addition to (1) the contingencies implicit in the externality of relations to impressions and ideas, our ways of thinking and acting are subject to a second order of contingency—namely, (2) the contingency of those relations that are at all affected (or effected) by society or social convention. As we will see, at least as far as understanding them theoretically goes, this includes nearly all of our perceptions.

We will also see, however, that despite its importance Deleuze's account fails to capture the complexity of the relationship Hume describes between nature and society in Hume's philosophy. In particular, Deleuze fails to consider just what for Hume makes society and custom themselves possible and how nature plays a role in their constitution. Like the individual imagination, the social and historical imagination, too, is powerful. But while the way it combines and separates impressions and ideas may be in many ways contingent, its operations, just like those of the individual imagination, involve natural forces of union and direction that guide and limit it, as well. Nature, in short, is not utterly marginalized by the contingencies of social convention. Indeed, in Hume's philosophy nature makes convention possible and convention completes nature. But this is not the end of the story.

In addition to failing to appreciate this complexity and to inquire about the conditions of the possibility of society per se, there's a still another, even deeper shortcoming to Deleuze's account, one he shares with many Anglo-American interpreters. Like many others across the Channel and across the sea, Deleuze fails to understand precisely what Hume means theoretically by "nature" and the "natural." Rather than undertake to consider reflexively what Hume's remarks about theory (what might be called his metatheory) reveal about the status and meaning of his own theoretical apparatus, interpreters have read Hume's use of "nature" in the best cases as consistent with usages common to early modernity and in the worst as a screen upon which to project their own contemporary prejudices.

By attending to Hume's metatheoretical remarks, however, we will see something else. We will see that in theorizing about "nature" and the "natural" Hume enters neither a metaphysical nor a realist claim. By that I mean that in deploying terms like "nature" and "natural" Hume pointedly refuses to make a claim about either the ultimate being or reality of things or that he has acquired knowledge of the world as it exists in itself, independently of our experience of it. While for Hume "nature" and the "natural" does—and rightly so, I think—point to those features of human existence that appear stable, necessary, and orderly to us, Hume is self-consciously aware that his discourse about nature and the natural is just that—a discourse. Hume's skeptical reflections lead him to a conclusion (or perhaps better, a refusal) he never abandons—that in philosophizing, even about nature, one ought not pretend to have discovered the essential nature of independent reality. One should refuse such pretension even while continuing to philosophize and to conduct natural science. In order to see this, let us begin by examining in more detail the interaction Hume portrays between society and nature in human convention.

II: Natural Conventions and Conventional Nature

There is a well-known paragraph describing the cooperation of two people rowing a boat that appears in Book III of Hume's *Treatise*.[4] The specific context of this passage is Hume's claim that the basis and origin of conventions—in this case the conventions

---

[4] Part II, Section II, "Of the origin of justice and property." The paragraph, in its entirety, runs as follows: "This convention [i.e. respect for possessions and property] is not of the nature of a *promise*: For even promises themselves, as we shall see afterwards, arise from human conventions. It is only a general sense of common interest; which sense all the members of the society express to one another, and which induces them to regulate their conduct by certain rules. I observe, that it will be for my interest to leave another in the possession of his goods, *provided* he will act in the same manner with regard to me. He is sensible of a like interest in the regulation of his conduct. When this common sense of interest is mutually express'd, and is known to both, it produces a suitable resolution and behaviour. And this may properly enough be call'd a convention or agreement betwixt us, tho' without the interposition of a promise; since the actions of each of us have a reference to those of the other, and are perform'd upon the supposition, that something is to be perform'd on the other part. Two men, who pull the oars of a boat, do it by an agreement or convention, tho' they have never given promises to each other. Nor is the rule concerning the stability of possession the less deriv'd from human conventions, that it arises gradually, and acquires force by a slow progression, and by our repeated experience of the inconveniences of transgressing it. On the contrary, this experience assures us still more, that the sense of interest has become common to all our fellows, and gives us a confidence of the future regularity of their conduct: And 'tis only on the expectation of this, that our moderation and abstinence are founded. In like manner are languages gradually establish'd by human conventions without any promise. In like manner do gold and silver become the common measures of exchange, and are esteem'd sufficient payment for what is of a hundred times their value" (*T* 490).

of private property and possession—are not those of formal contract (of, that is, verbalized agreement and commitment expressed in speech or writing). Earlier political philosophers like Thomas Hobbes (1588–1679) and John Locke (1632–1704) had argued that the legitimacy of the state is grounded in a contractual agreement entered by the members of the social order it governs.

Hume, however, took a different view, finding prevailing social contract theory philosophically baseless and the politics it engendered not only distasteful but also, in too many cases, downright dangerous. For Hume, the notion of contract presents an inadequate account of the bases of society and the legitimacy of the state. Contractual agreements are themselves only possible through the existence of something deeper, namely the unreflective agreements and alignments of human nature and sensibility. Contracts and promises are, therefore, for Hume, derivative phenomena. Contrary to the claims of social contract philosophers, they only "arise from" prior "human conventions" and as such are insufficient, by themselves, to ground the political institutions social contract philosophers wished to legitimate with them.[5]

A great deal, then, is at stake for Hume in his thoughts on convention. In the form of what he calls "general rules," conventions bring structure and order to society by underwriting social institutions such as the law, the state, family, and experimental science. General rules, may even affect perception itself: "our adherence to *general rules*…has such a mighty influence on the actions and understanding" that it "is able to impose on the very senses" (*T* 374). General rules articulate, modify, and channel sympathies and passions, thereby developing and directing moral and aesthetic sensibilities. Hume acknowledges this in the provocative text, "A Dialogue," he attaches to the *Enquiry concerning the Principles of Morals* (1751).[6] In the course of the dialogue,

---

[5] See Hume's "Of the Original Contract" in *Essays: Moral, Political, and Literary*, ed. Eugene F. Miller (Indianapolis: Liberty Classics, 1985), 465 ff. See also *T* 516–25.
[6] David Hume, *Enquiry concerning the Principles of Morals* in *Enquiries concerning Human Understanding and concerning the Principles of Morals*, edited by L. A. Selby-Bigge and P. H.

Hume maintains that what eighteenth-century Britons would count as parricide and incest passed readily among the ancient Romans as acceptable, honorable, and even desirable behavior. Indeed, "We can form no wish, which has not a reference to society" (*T* 363). Convention structures our ways of thinking, perceiving, and doing.

So pervasive and important, in fact, are general rules, that without them, implies Hume, human selves would lack integrity and human society would fail to achieve the pleasant stability and consistency necessary for its coherence and continued existence. Without conventions, therefore, neither human selves nor the social orders that they compose would be possible. Convention is a necessary condition for the self, for science, for society, and, in a sense, for the world as we know it. When Hume writes about "convention," then, he is literally writing about the convening or coming together of people, about their facing and in some way surmounting in sustainable ways the otherness separating them, the gap of difference which skepticism exploits. If one is to look for Hume's reflections on skepticism with regard to other minds (as opposed to skepticism with regard to the external world) it is here, in his thoughts on convention, to which one should turn.

Although his "Abstract" or "Advertisement" might seem to suggest otherwise, Hume's remarks on general rules in Book III of the *Treatise* are not, therefore, extraneous to the philosophical project he had taken up earlier in Books I & II.[7] Deleuze is right that the principles of human nature and the natural relations of ideas Hume articulates are not—understood simply in terms of individual minds—sufficient to provide an understanding of human beings as they exist in society. Human thoughts and feelings are not, for Hume, formed simply through the confrontation of the individual mind with the world in individual experience. Rather, thoughts and feelings are formed through the confrontation of social selves with the world—and with each

---

Nidditch, 3rd edition (Oxford: Clarendon Press, 1975). Hereafter *EPM* and *EHU*.

[7] The "Advertisement" describes it as a text that can be considered independent of the first two Books. Of course, this is not to say that it must be considered as such.

other.[8]   But just how is human convention itself possible for Hume? What has Deleuze missed?

The genesis of "observations," "expectations," and calculations constitutive of convention formation involves and requires, for Hume, much more than arbitrary choice or immediate, innate, "hardwiring," simple-minded social construction, authority, or philosophical demonstration. Rather, conventions are generated principally through the complex, educating, natural, and social processes Hume calls "experience." This process does not itself take place on the basis of explicit contract or agreement but only "arises gradually," acquiring "force by a slow progression."

The experiences relevant to, for example, the conventions of justice include the experience of "inconveniences," such as the physical pain accompanied by the loss (or threat of loss) of one's own possessions likely to follow upon one's interference with another's. If I am right to discern a confrontation with skepticism animating Hume's thoughts on convention, then his referring to "inconveniences" may be read literally and thereby call our attention to something of the nature of convention itself. What I wish to suggest is that Hume's mention of inconvenience as a crucial factor in the formation of convention resonates with a sense of those occasions when we are not able to convene with one another, when our desires and hopes and expectations of what others will or ought to do are thwarted, frustrated, or disappointed—when we are left at a loss in the face of others as other.

It is in attempting to overcome this otherness that humans produce convention: but Hume argues that such overcoming is not simply a matter of calculative agreements. It also depends, more deeply, upon (a) certain possibilities we possess and upon (b) a kind of sentimental concord, what he calls in the paragraph to which I have referred a "general sense" of "common" interest (T 490). This sort of sensibility may be the result but not the creation of deliberative calculation. Not an arbitrary invention and not

---

[8] See Deleuze, *Empiricisme et subjectivité* (1953), especially ch. 2, "Cultural World and General Rules."

determined according to formal agreements such as contracts or promises (Hobbes) or divine ordinations or some sort of deus ex machina (Descartes, Anselm, Augustine), such a sensibility must rather for Hume be rooted in our common, human life and human nature. In this sense, convention seems to require a natural basis.

Moreover, while common sensibility is modified, developed, and diversified through the course of contingent culture and history, no interest *could* become common, no culture and history *could* become operant, and no fellow-feeling *could* be produced without the antecedent, common capacities for achieving sentimental alignment that seem natural and inseparable from us, to which we seem fated.[9] Like the "secret union" of association that sets a check to the empire of individual imagination, here too Hume signals that amidst the contingency and artifice of convention formation we find ourselves depending upon the common and the natural to bring us together so that we may engage a shared world of meaning and practice.

In Hume's image, two people work together rowing a boat across a body of water. In such a context, they "know" each other, their individual interests as well as their common interest in this shared situation; they possess and act upon expectations of causal sequence, behavior, and character. They have reached a common understanding of their situation through their shared work, through their common experience of the nature and purpose of boats, of travel across water, of the structures of play and transportation and time in their culture, and perhaps through their own

---

[9] Plato's Socrates offers a remark which is consonant with Hume's sense of what transpires in convention/community formation: "in our community, then, above all others, when things go well or ill with any individual everyone will use that word 'mine' in the same sense and say that all is going well or ill with him and his....And, as we said, this way of speaking and thinking goes with fellow-feeling; so that our citizens, sharing as they do in a common interest which each will call his own, will have all their feelings of pleasure or pain in common" (*Republic*, V 463c–464a, trans. Cornford). This image may also be taken as an extreme form of the sympathetic identification of one's own interest with the interests of the community which constitutes Hume's reply to Hobbes, Mandeville, and other neo-Augustinian moralists. Plato's Socrates sees in this communion "the greatest good." (See also *Republic*, III 412b.)

interpersonal histories, as well. In short they have achieved and depend upon what Hume calls a "common life."

For Hume this sort of activity—whether or not grounded in explicit agreement—depends upon the prior condition of our naturally *being able* to row together—that is, upon the intrinsic, human *possibility* of engaging in a world of common interest and practice. The existence of this possibility, in turn, depends for each thinker upon its being constitutive of our *nature*, of our natural capacity for having certain shared feelings and desires, for having complementary expectations that occurrences have consequences, for having mutually comprehensible pleasures and pains, attractions, interests and revulsions. That the actual achievement of fellow feeling may be the result of *recognizing* (that is, understanding) common interests does not render sentiment posterior to calculation, for no interest could produce fellow feeling were not the capacity for it already present; and no calculations could take place without the prior sentimental alignments which underwrite the institutions of rationality.

There is more. In discussing the conventions of justice, Hume maintains that nature completes itself in convention.[10] The development of rules governing the stability and transference of property (*T* 256) does not, for Hume, entail the imposition of alien artifice upon the natural other. Rather, for Hume, the progress of convention (and with it the progress of sentiment) marks a change in nature itself, a change from an "*uncultivated*" (*T* 488) to a "methodized" and "corrected" nature (*EHU* 130 [162]). Hume depicts this continuity between the conventional and the natural more emphatically when he describes the achievement of certain "fundamental" (*T* 526) conventions as, in fact, the invention of new "laws of nature."

---

[10] About the partiality and selfishness of humanity, Hume writes, for example: "The remedy, then, is not deriv'd from nature, but from *artifice*; or more properly speaking, nature provides a remedy in the judgment and understanding, for what is irregular and incommodious in the affections" (*T* 489).

Tho' the rules of justice be *artificial*, they are not *arbitrary*. Nor is the expression improper to call them *Laws of Nature*; if by natural we understand what is common to any species, or even if we confine it to mean what is inseparable from the species" (*T* 484).[11]

One might, therefore, upon Humean lines, distinguish (1) basic human nature from (2) artificial but nonarbitrary "laws of nature" and contrast these two against a third type: (3) more contingent arrangements made by human society. It is one thing, for example, to produce different conventions governing the proper manner of laying out silverware (perhaps, e.g., laying the knife at the bottom of the plate rather than to the right side). It is another to create different prohibitions governing sexual relationships and private property. It would be something else, yet again, as Stanley Cavell points out, to expect that our actions will not have consequences in the world, to count entirely different expressions (and not those we now count) as expressions of pain, or to "be bored by an earthquake or by the death of [one's] child or the declaration of martial law, or...quietly (but comfortably?) sit on a chair of nails."[12] Concerning such fantastic revisions of human convention, Cavell observes:

That human beings on the whole do not respond in these ways is, therefore, seriously referred to as conventional; but now we are thinking of convention not as the

---

[11] Heraclitus is recorded as having said: "Man's *ethos* [character] is his *daimon* [fate, spirit, divinity, fortune]" (CXIV, D.19). See Charles H. Kahn, *The Art and Thought of Heraclitus: An Edition of the Fragments and Commentary* (Cambridge: Cambridge University Press, 1979). See also Martin Heidegger's comments on this passage in his 1947 "Letter on Humanism." The Latin *natura* is related to *nasci*, to be born, and hence "nature" bears the sense of what one is born to, what is natal, i.e., one's fate.

[12] Stanley Cavell, *The Claim of Reason: Wittgenstein, Skepticism, Morality, and Tragedy* (Oxford: Clarendon Press, 1980), 111; hereafter, *CR*. I also take it that this is the type of convention or agreement to which Ludwig Wittgenstein (*Philosophical Investigations*, trans. G. E. M. Anscombe [Oxford: Basil Blackwell, 1953], hereafter *PI*) refers to at *PI* #355: "the point here is not that our sense-impressions can lie, but that we understand their language. (And this language like any other is founded on convention [*Übereinkunft*].)"

arrangements a particular culture has found convenient, in terms of its history and geography, for effecting the necessities of human existence, but as those forms of life which are normal to any group of creatures we call human, any group about which we will say, for example, that they *have* a past to which they respond, or a geographical environment which they manipulate or exploit in certain ways for certain humanly comprehensible motives. Here the array of "conventions" are not patterns of life which differentiate human beings from one another, but those exigencies of conduct and feeling which all humans share. (*CR* 111)

There are, then, according to both Hume and Cavell, different orders of convention. Some we might characterize as "basic" or fundamental, others as relatively "superficial" or "surface" agreements.[13] I should like to say that those agreements or conventions that are basic are also "essential" in the sense of being definitive of human ways of being, definitive such that they could not be altered without altering the very character of what we recognize or what appears to us as human life.[14]

Surface or accidental conventions by contrast are agreements or alignments that are revisable, or at least revisable in ways that will not fundamentally disrupt what we recognize as or what appears to us as our ways of being. What is crucial, then, is not whether or not such conventions are grounded in metaphysical essences or even in an external natural order that reason or perception can grasp. The key, rather, is resistance—perhaps what a Sartrean might describe in terms of a "coefficient of adversity"—the extent to which we find and recognize that certain practices or beliefs are not plastic, revisable, or open to revision.

---

[13] Echoing what Wittgenstein calls "surface grammar"(*PI* #664); a notion related to Wittgenstein's search for fundamental or "grammatical" propositions. Wittgenstein, *PI*, ##232, 251, 293, 295, 371, 373, 458, 496, 497. "Essence is expressed by grammar" (*PI* #371).
[14] The *OED* tells us that the Latin *natura*, from which "nature" is derived, refers to the character, constitution, or course of things.

Along just these lines, for Hume the "natural" signals the easy, stable, common, not easily revisable and regular features we recognize of the world and ourselves, and what is natural to us can at least in part be artificial.[15] This way of conceiving the natural seems to jibe well with the "mitigated" or "academical" skepticism to which Hume returns after nature breaks his excessive, solitary, and uneasy skeptical doubts insofar it is characterized as "durable and useful" (*EHU* 161 [129]). But how, more precisely, does this conception of nature and the natural relate to Hume's skepticism?

III: Natural Science: Ancient, Modern, and Skeptical

It is crucial to see that for Hume even the most basic features of the world and ourselves that appear as natural to us remain fundamentally fragile and vulnerable to revision and to skeptical doubt—contingencies marshaled by the empire of imagination. Perhaps most famously in this regard, Hume calls into doubt our reasons for believing in the existence of the causal connection and the uniform operations of nature across time—for example, the sun's rising or not rising tomorrow (*T* 77, 89, 124).[16] Because of this, when formulating his view of natural science and this understanding of nature, Hume departs from dominant conceptions of scientific law available in his time.

Of course, Hume rejects what he understands to be the project of ancient philosophy and science. Like Jean Buridan (fl. 14th century), Hume defines "ancient" philosophy as that which

---

[15] For more on the meanings of "nature" as Hume uses it, see Miriam McCormick, "Hume on Natural Belief and Original Principles" *Hume Studies* 19.1 (April 1993): 103–16.

[16] It is possible Hume was inspired to consider this question through encountering it in discussions taking place among the medical and scientific community of Edinburgh of his day. See Craig Walton, "Why Do We Believe the Sun Will Rise Tomorrow? Dr. George Young's Medical Lectures of 1730–31 and the Young David Hume." An unpublished paper. Las Vegas, Nevada: Department of Philosophy and Institute for Ethics and Policy Studies, University of Nevada at Las Vegas, 1990.

maintained the reality of substantial forms (*T* 221), and the apprehension of form through reason may be understood as the project of Platonic and Aristotelian science.[17] According to these ancient traditions, independent reality is grounded in immaterial forms that give to things their specific unity and identity through qualitative change. Substantial forms structure rude matter into kinds or types of entities and thereby endow them with being and intelligibility.

For Hume substantial forms are unintelligible. To the extent we have an idea of them at all we do so only through the activities of fanciful imagination and the habits our minds develop in perceiving the way things change in continuous and gradual ways. Hume writes: "When we gradually follow an object in its successive changes, the smooth progress of the thought makes us ascribe an identity to the succession; because 'tis by a similar act of the mind we consider an unchangeable object" (*T* 220). If the term "substance" is taken as referring to some sort of metaphysical ground or substrate and not to some continuous collection of perceptions, it is literally senseless. In a strictly metaphysical sense, we have no idea of transcendent reality at all. Metaphysicians' usage of terms like substance, for example, is a linguistic deception made possible by the abbreviated way we often use meaningful terms:

> For it being usual, after the frequent use of terms, which
> are really significant and intelligible, to omit the idea,
> which we wou'd express by them, and to preserve only
> the custom by which we recal the idea at pleasure; so it
> naturally happens that after the frequent use of terms,
> which are wholly insignificant and unintelligible, we
> fancy them to be on the same footing with the prece-
> dent, and to have a secret meaning, which we might
> discover by reflection. The resemblance of their appear-

---

[17] Jules Michelet, writing in the nineteenth century, would solidify the modern distinction between the ancient, medieval, and modern historical periods; *Introduction à l'histoire universelle* (1831).

ance deceives the mind, as is usual, and makes us imagine a thorough resemblance and conformity. By this means the philosophers set themselves at ease, and arrive at last, by an illusion, at the same indifference, which the people attain by their stupidity, and true philosophers by their scepticism. (*T* 224)

Hume's rejection of ancient and medieval metaphysics is unsurprising. But Hume also explicitly rejects what he regards as "modern" philosophy. For Hume, "modern" philosophy is a species of thought that originated with thinkers like Pierre Gassendi and Galileo and traversed a course through Descartes, Malebranche, Newton, and Locke.[18] It is that species of thought that divides the perceptual order into two components, those of "primary qualities" and "secondary qualities." Secondary qualities (e.g., color, taste, smell, and texture) are those features of perceptions that are held to have no independent existence beyond our perception of them; they exist entirely "within" the "mind." Primary qualities (viz. extension and solidity) are said to correlate, tightly or loosely, with independently existing entities.   They are eminently quantifiable, and as such primary qualities legitimated the claim of the new, modern, mathematical science to disclose the nature of independent reality.

Hume criticizes this form of philosophy on grounds analogous to those upon which he criticizes ancient philosophy. Just as the supposition of imperceptible substantial forms is for Hume baseless and without philosophical warrant, so too is the supposition that primary qualities correlate with independent and imperceptible entities while secondary qualities do not.   It is impossible, according to Hume, to conceive of solidity without color. It is impossible to conceive of extension without texture.

[18] Galileo Galilei, *Dialogue concerning the Two Chief World Systems: Ptolemaic & Copernican* (1629); René Descartes, *Meditations on First Philosophy* (1640) and *Principles of Philosophy* (1644); Pierre Gassendi , *Syntagma Philosophicum* (1659); Nicolas Malebranche, *Recherche de la verité* (1674–75); Isaac Newton, *Philosophiae Naturalis Principia Mathematica* (1667); John Locke, *An Essay concerning Human Understanding* (1689).

In short, it is impossible to conceive of primary qualities independent of secondary qualities. Upon the whole, writes Hume, we "must conclude, that after the exclusion of colours, sounds, heat and cold from the rank of external existences, there remains nothing, which can afford us a just and consistent idea of body" (*T* 229). "Our modern philosophy, therefore," says Hume, "leaves us no just nor satisfactory idea of solidity; nor consequently of matter" (*T* 229). We must read, then, Hume's apparently Newtonian subtitle of the *Treatise* ("*An Attempt to introduce the experimental Method of Reasoning in Moral Subjects*") as at least in part ironic.[19]

Hume's criticism of what he takes to be both ancient and modern thought, therefore, goes beyond arguing that it is false or inconsistent. Following Berkeley, Hume argues that the very terms and sentences which ancient and modern philosophy deploys in its theories are literally meaningless.[20] In this way Hume's thought may be aligned with work by more recent analytic philosophers who have tried to show the literal senselessness of discourses purporting to disclose any reality independent of what we perceive.[21]

Hume's scrutiny not only, however, undermines modern philosophy's claims to knowledge of external, extra-perceptual posits. Hume surpasses Berkeley in the very next section of the *Treatise* (Pt. IV, §V) by also subverting the doctrine of an immaterial soul, a dogma particularly dear to the bishop. In the well-known §VI of Part IV of the *Treatise*, Hume also addresses the substantial "self" said to remain constant through the course of "internal" experience, replacing it with the commonly-called "bundle" theory of the self. According to the bundle theory, the self is a composite, a continuously changing aggregate, with no non-arbitrary principle of order and no fundamental "I" beyond

---

[19] For a rather different view on Hume and Newtonianism, see Nicholas Capaldi, *David Hume: The Newtonian Philosopher* (Boston: Twayne, 1975).
[20] Berkeley had leveled similar criticisms of materialism in both his *A Treatise concerning the Principles of Human Knowledge I* (1710) and *Three Dialogues between Hylas and Philonous* (1713).
[21] For example consider Wittgenstein, Hilary Putnam, Richard Rorty, James Conant and even Immanuel Kant.

the accidental collocation of perceptions composing any particular set of experiences.[22] The self, in other words for Hume is an event, not an entity.[23] "Our self, independent of the perception of every other object, is in reality nothing" (*T* 340).

Finally, consider how Hume's conception of the laws of nature differs from those advanced by rationalistic philosophers and natural theologians. For Hume certainty about the necessary causal connections described by natural laws depends only upon the human mind and its engagement with experience. Hume famously shows against rationalists like Descartes and Clarke that experience gives no evidence of any logical or intrinsic connection between causes and their effects. All we experience are constant conjunctions and a feeling of necessity within our own breasts (*T* 73ff.).

To see Hume's view more clearly, consider an analogy.[24] Grammarians discern grammatical rules describing the practices of language. The practices of language were not originally produced by those rules. Language developed first; the rules were determined later. Of course, in particular cases, the grammatical rules can subsequently be used to manipulate and discipline linguistic practices. Generally speaking, however, no one in particular invents language, and no one has the power to change it. To the

---

[22] It is worth mentioning, however, that Hume was not entirely comfortable with his bundle theory of the self. In an appendix Hume attached to the *Treatise* he confessed that: "I had entertain'd some hopes, that however deficient our theory of the intellectual world might be, it wou'd be free from those contradictions, and absurdities, which seem to attend every explication, that human reason can give of the material world. But upon a more strict review of the section concerning *personal identity*, I find myself involv'd in such a labyrinth, that, I must confess, I neither know how to correct my former opinions, nor how to render them consistent. If this be not a good *general* reason for scepticism, 'tis at least a sufficient one...for me to entertain a diffidence and modesty in all my decisions" (*T* 633ff.).

[23] Hume writes: "setting aside some metaphysicians...I may venture to affirm of the rest of mankind, that they are nothing but a bundle or collection of different perceptions, which succeed each other with an inconceivable rapidity, and are in a perpetual flux and movement....There is properly no *simplicity* in it at one time, nor *identity* in different; whatever natural propension we may have to imagine that simplicity and identity" (*T* 253).

[24] I am extremely grateful to Prof. Rollin Workman of the Philosophy Department of the University of Cincinnati for offering me the core of this analogy as a way of clarifying my thoughts on this matter.

extent (a) grammar can be revised the rules are contingent, and to the extent they are (b) created through human artifice the rules are conventional. But to the extent (a) none of us created and none of us can change linguistic practices and to the extent (b) those practices are grounded in the possibilities of human nature, the rules are not contingent.

From the point of view of one we might call a grammatical Platonist or realist, there is an independent, external grammar, which it is the job of grammarians to discern and represent in their theories. For a Humean grammarian, by contrast, it seems impossible to know whether such a Platonic or independent grammar exists. Moreover, we are probably better off not pretending to having apprehended it.

Now, substitute "nature" for "practices of language" and "natural law" for "grammatical rules." Accordingly, for Hume nature is not produced by natural laws—though we can manipulate and even complete nature through them. Natural laws are contingent and conventional to the extent that we formulate them on the basis of empirical data we collect and we organize and to the extent we are able to revise those laws. Natural laws are, on the other hand, not contingent to the extent that we use them to describe regularities of the world we experience that are beyond our control and not our invention. They are also natural to us in the sense that the very possibility of their formulation exhibits possibilities of our way of being. For the Humean skeptic, while the natural laws we formulate may be useful and durable and satisfying to us, we should not pretend that they represent an independent or a metaphysical order.

Rationalists, deists, and other naturalistic philosophers, however, like Platonic grammarians hold that the laws they formulate and the philosophical theories they develop go much further. Rationalists like Descartes, Spinoza, and Leibniz—as well as other realists—maintain that their reasonings disclose a logical, intrinsic and metaphysical connection between cause and effect. Deists ground the connection in the divine, a divine realm to which they speciously maintain they have had access. Natural laws, say deistic natural theologians, are divine commands. Discovery of such laws requires, therefore, the acquisition of metaphysical knowledge, and the order and harmony of those laws is grounded in the order, harmony, and rectitude of God.[25] The necessities described by natural laws are rooted in the unchanging nature and power of God, and one's certainty in that necessity is established by deductive reason. Natural science is, therefore, for the deists and other natural theologians a species of natural religion. Tom Paine, for example, refers to the discovery of the principles of natural science—which was then called natural philosophy—as "the true theology."[26] Indeed, for Paine the only true Scripture is the "Bible of Creation." "THE WORD OF GOD," he writes, "IS THE CREATION WE BEHOLD" (*Age of Reason*, 482). We can see, then, that Hume's skeptical conception of science and scientific law differs both from those based upon ancient and medieval notions of form as well as from modern efforts to determine the world revealed through primary qualities and deductive reason.

The rediscovery of ancient skeptical texts by European thinkers in the fifteenth century provided thinkers of the time with conceptual instruments with which to dismantle scholastic thought; it also, however, precipitated an intellectual crisis as philosophers sought ways to answer skeptical challenges and place knowledge claims on a new foundation. Indeed, the project of modern philosophy may in large measure be understood as the project of trying to overcome the challenge of skepticism. Among the dominant movements in the modern struggle against skepticism must be counted rationalism, primary quality empiricism, and corpuscularism, as well as natural and revealed theology. Hume's embrace with skepticism, therefore, signals his disengage-

---

[25] On the deists' conception of God consult Étienne Gilson, *God and Philosophy* (New Haven: Yale University Press, 1941), 104ff; especially Chapter III, "God and Modern Philosophy," 74–108. And see my "Hume, Skepticism, and Early American Deism," *Hume Studies* 25.1&2 (November 1999): 171–92; where I discuss this issue at greater length.

[26] Thomas Paine, *The Age of Reason*, in *The Complete Writings of Thomas Paine*, edited by Philip S. Foner (New York: Citadel Press, 1945), I:487–88.

ment from the modern project; it is an embrace which brought down barrages of criticism and purple steam upon him and contributed to his twice failing to acquire an academic post—once at the University of Glasgow and once at the University of Edinburgh. Perhaps the most dramatic of such invective was produced by James Beattie, who in his "Castle of Skepticism" (1767) characterized Hume as the devious inhabitant of a dark fortress, bent upon luring unsuspecting travelers inside and then subjecting them to inhuman tortures.

For Hume, however, just the inverse is true. It is not skepticism but, rather, the misguided attempts to overcome skepticism (to establish divine or rationalistic foundations for science and philosophy) that leads to suffering. Hume warns against these pathologies, and others, by comparing them to the torments of Tantalus and by characterizing them as what he calls "false philosophy" (T 223).[27] Theories and laws of philosophy and natural science are in Hume's view best conceived as the products of human convention, custom, habit, invention, and imagination rather than as attempts to mirror the nature of an independent natural world. Again, if there is such a nature, then for Hume it appears we have no way of knowing whether our beliefs correspond to it, and it seems happier for us not to pretend that they do.

Certainly, within Hume's theoretical apparatus we find him appealing to human nature, natural laws, and natural principles. But as we have just seen, Hume distances himself from what he thinks of as the modern approach to philosophy and science, the philosophical project of overcoming skepticism and discerning features of an independently existing world. As we are about to see in the next section, we also find in his work metatheoretical

---

[27] But is it anything more than an empirical—and therefore merely probable—claim that Tantalus will never reach the fruit he desires, or, for that matter, that Sisyphus will never succeed in raising the rock to a permanent place on the top of the hill? What grounds do we have for thinking so? For an expansive articulation of Hume's theory of true and false philosophy see Donald W. Livingston, *Philosophical Melancholy and Delirium: Hume's Pathology of Philosophy* (Chicago: University of Chicago Press, 1998).

reflections through which Hume makes abundantly clear that *nothing* of his philosophical system is to be construed as transcending, rebutting, or defeating his skepticism.[28] For Hume science and philosophy—indeed, all of human theoretical life—is rife with the contingency not simply of social convention but of skeptical doubt and suspension. Philosophy for Hume acknowledges human finitude.

IV: Implying "no dogmatical spirit"

Keep sober and remember to be skeptical.[29]
— Epicharmus

What's important to our purpose here is understanding that Hume's thoughts about nature—like his thoughts about custom and tradition—are always and already wrapped in a skeptical sense of their provisional, contingent, fragile character. As such they occupy a categorically different status from the contentions about nature made by dogmatists like the Lockeans, Newtonians, Cartesians, deists, and Common Sense theorists—all of whom make claim to various kinds of realism for their theories, realisms that Hume, as a skeptic, will not endorse. Nor will Hume, on similar grounds, endorse even a Berkelean idealism. Berkeley,

---

[28] This is not to say Hume's view of natural law is utterly different from conceptions developed more recently. The claims Hume associates with natural law are true, non-analytic, universal generalizations, whose subject terms are unrestricted, which sustain counter-factual conditionals, and which may be used to formulate explanations and predictions of events in nature. These features of Hume's claims are standard characteristics of natural laws that have been developed by regularist philosophers of science. See: A. J. Ayer, "What is a Law of Nature?" *Revue internationale de philosophie* 36 (1956); Karl R. Popper, *The Logic of Scientific Discovery* (London: Hutchinson, 1959), ch. 3. Other supporters of this view include Carl Hempel, J. L. Mackie, Ernest Nagel, Arthur Pap, and Bertrand Russell. One may wish to add "expressing a necessary relation" to the list of features characteristic of law-like statements. Hume would, no doubt, assent.

[29] On the back of one of the sheets of his memoranda Hume wrote this remark in the original Greek; see Ernest Campbell Mossner, "Hume's Early Memoranda, 1729–40: The Complete Text," *Journal of the History of Ideas* 9.4 (1948): 503n17.

despite his rigorous empiricist criticisms on the meaningfulness of the concept of material substance, unabashedly roots perception in the agency of God acting on spirit. For Hume, by contrast, the ultimate source and cause of impressions is something about which he simply maintains a bracketed silence—or, as the ancient Pyrrohnians would call it, *aphasia*.[30] Hume's approach to nature and natural law in this regard is not only distinct, as we have seen from the ancient and modern projects of philosophy; it is also consistently and profoundly skeptical. To better understand just how skeptical, let us revisit one of the most important sources for the transmission of skeptical ideas to modernity.[31]

In the *Outlines of Pyrrhonism*, Sextus Empiricus characterizes the skeptical *agoge* (or way of life) this way:

> Adhering, then, to appearances we live in accordance with the normal rules of life, undogmatically, seeing that we cannot remain wholly inactive. And it would seem that this regulation of life is fourfold, and that one part of it lies in the guidance of Nature, another in the constraint of the passions, another in the tradition of laws and customs, another in the instruction of the arts. Nature's guidance is that by which we are naturally capable of sensation and thought; constraint of the

---

[30] See *Treatise* 2n1: "I here make use of these terms, *impression and idea*, in a sense different from what is usual, and I hope this liberty will be allowed me. Perhaps I rather restore the word, idea, to its original sense, from which Mr. *Locke* had perverted it, in making it stand for all our perceptions. By the term of impression I would not be understood to express the manner, in which our lively perceptions are produced in the soul, but merely the perceptions themselves; for which there is no particular name either in the *English* or any other language, that I know of." Like the ancient Pyrrhonists who preceded him and the phenomenologists who would later follow him, Hume endeavors to stick with appearances (*phainomena*) and resist speculating about unobservable entities, such as substantial forms or material objects, that purportedly either cause or order them. Concerning Hume's relationship to phenomenology see George Davie, "Edmund Husserl and 'the as yet, in its most important respect, unrecognized greatness of Hume,'" in *David Hume: Bicentenary Papers*, ed. G. P. Morice (Edinburgh: Edinburgh University Press, 1977).

[31] Two of the others are Cicero's *De Natura Deorum* and *Academica*. About the influence of these on Hume, see my "Doubt and Divinity: Cicero's Influence on Hume's Religious Skepticism," *Hume Studies* 20.1 (April, 1994), 103–20.

passions is that whereby hunger drives us to food and thirst to drink; tradition of customs and laws, that whereby we regard piety in the conduct of life as good, but impiety as evil; instruction of the arts, that whereby we are not inactive in such arts as we adopt. But we make all these statements undogmatically. (*PH* I:23–24, Chapter XI)[32]

Although Sextus does little in the way of explicitly explicating or elaborating upon it, this is no insignificant passage in his work. Brief though it may be, this passage describes nothing less than what Sextus calls the "criterion" of skepticism—the criterion for thinking and acting to be adopted upon engaging skeptical life. What I find so interesting is that—with the exception of the endorsement of piety—so much of this passage is so very Humean: the deference to nature and custom, the constraint of the passions, the acknowledgment of the importance of practical arts, the abjuring of dogmatism. The spirit of the passage is so similar to Hume's that one may easily wonder as to whether good David might have read and absorbed it.[33]

Whether he actually read the passage or not, however, is of little philosophical import. In either case, it remains true that Hume's own metaremarks on the standing of his philosophical (and nonphilosophical) assertions follow along just the sort of lines described by Sextus. Among the most telling are these words with which Hume closes the whole of Book I of the *Treatise*, proclaiming that he, like Sextus, "makes all" his "statements undogmatically":

---

[32] Translation Bury. Sextus's *Outlines of Pyrrhonism* (hereafter also "*PH*") is composed of three books, the first outlining the skeptics' views, the latter two attacking various forms of what appeared to them as "dogmatism." R. G. Bury, *Sextus Empiricus*, 4 vols. (Cambridge: Harvard University Press, 1933).

[33] See my "The Bibliographic Bases of Hume's Understanding of Sextus Empiricus and Pyrrhonism" *Journal of the History of Philosophy* 36 (1998): 261–78.

> I here enter a *caveat* against any objections, which may
> be offer'd on that head [i.e., the charge of dogmatism];
> and declare that such expressions were extorted from me
> by the present view of the object, and imply no dogmat-
> ical spirit, nor conceited idea of my own judgment,
> which are sentiments that I am sensible can become no
> body, and a sceptic still less than any other.   (*T* 274)

The eminent Hume scholar Richard H. Popkin is therefore in
error when he writes: "if one is really Pyrrhonian, as Hume was,
one will be as dogmatic and as opinionated as one is naturally
inclined to be."[34]   The notion that Hume embraces dogmatic
philosophy is utterly contradicted by this passage. Not only does
Hume explicitly reject dogmatism and a naturalistic basis to
skepticism. Hume actually uses skepticism to order and refine just
what ought properly to be by meant by "nature" and the "natural."
What ought properly to be meant is a non-dogmatic concept of
both. Popkin, in short, like many others, has put the cart before
the horse, for it is not for Hume "nature" that justifies engaging in
skeptical philosophy; rather, it is skeptical philosophizing that
defines the sort of "nature" that makes a proper sort of naturalism
possible.

We might say, then, that Part IV of Book I is the true
philosophical beginning of Hume's *Treatise*, for it is from the
peculiar vantage point Hume reaches there, after struggling with
skeptical arguments and acknowledging their implications, that he
then goes on to develop his philosophical theories. It is a philosophical standpoint from which Hume has come to acknowledge
the finitude of theory and yet has chosen nevertheless to go on
theorizing. "Methinks," Hume writes:

> I am like a man, who having struck on many shoals, and
> having narrowly escap'd ship-wreck in passing a small

---

[34] Richard H. Popkin, "David Hume: His Pyrrhonism and his Critique of Pyrrhonism" *The
Philosophical Quarterly* 1.5 (1951), 406.

frith, has yet the temerity to put out to sea in the same leaky weather-beaten vessel, and even carries his ambition so far as to think of compassing the globe under these disadvantageous circumstances. (*T* 263–64)

Popkin would have us believe that Hume's willingness to embrace philosophical theory despite having failed to expurgate skepticism is a sign of his willingness to be "dogmatic"—just as dogmatic as any other philosopher—when the "mood" strikes him to be so. William E. Morris would have us believe that the voice of this passage is not Hume's at all but some imagined metaphysician. Because Morris altogether fails not only to grasp the possibility of nondogmatic philosophy but also to acknowledge Hume's skepticism, it seems to him a "complete mystery" that Hume should go on philosophizing and accept the skeptical conclusions.[35] As we have seen, however, Hume's text is clear that although he continues to philosophize—to make assertions, to form concepts, to develop reflectively theories and beliefs, even about nature—he does so without any "dogmatic spirit." That is to say, the sort of theorizing Hume engages is a *different* sort of theorizing than that of the dogmatist. As his conception of "nature" is a central element of his philosophical-theoretical apparatus, it too must be presented in a manner *different* from the manner in which dogmatists of his time (and ours) present their naturalisms. Hume distills his thoughts on the standing of theory this way:

we might hope to establish a system or set of opinions, which if not true (for, perhaps, that is too much to be hop'd for) might at least be satisfactory to the human mind, and might stand the test of the most critical examination. (*T* 272)[36]

---

[35] William Edward Morris, "Hume's Conclusion" *Philosophical Studies* 99 (2000): 96.
[36] In the *Enquiry concerning Human Understanding*, Hume describes his skeptical mode of philosophy as "durable" and "useful" rather than true in the sense of accurately corresponding to some independent reality (*EHU* 161 [129]; Section XII, Part III).

A short while later, in a passage closely tied to Hume's advocacy of the "science of man," he continues describing this manner of skeptical theorizing:

> The conduct of a man, who studies philosophy in this careless manner, is more truly sceptical than that of one, who feeling in himself an inclination to it, is yet so over-whelm'd with doubts and scruples, as totally to reject it. (*T* 273)

Returning to common life and engaging in naturalistic theorizing dispels Hume's skeptical doubt, but it does not dispel his skepticism. He remains "truly sceptical." Hume's conception of "nature" does not represent his overcoming or lapse from skepticism; rather, it is itself skeptical or, perhaps more accurately, is alloyed with skepticism. While it would be wrong, therefore, to say that Hume wishes to take his claims dogmatically, it would not be wrong to take his contentions (including those about nature) philosophically and seriously.

V: Science, Politics, and Contingency

Confirmation for this reading may be found in Hume's political philosophy, for we find there a strategy for dealing with the forces of contingency analogous to his approach to contingency in natural science. To begin with, consider that for Hume it is not simply that conventional society determines the varieties and contingencies of natural passion; passion, deeply rooted in nature, turns round and imbues society itself with instability and fragility. Indeed, Hume often seems more concerned with the socially disruptive potential of the theories and practices he criticizes than with logical issues of consistency or veracity. Dogmatic religion and other superstitions are to be condemned not merely because their claims are baseless and their reasonings specious. "Generally speaking," Hume reminds us, "the errors in religion are dangerous;

those in philosophy only ridiculous" (*T* 272). Religion is not merely erroneous, it is a species of what Hume, and many others of the time, call "enthusiasm."

Hume's social-political philosophy reflects his acknowledgment of the pervasive and continuously disruptive potential of passion in the social order. Contrary to the claims of Thomas Jefferson, Hume's criticism of John Wilkes (riots in 1763) and the Whigs was not an expression of his philosophical sympathy for monarchy. It was, rather—much in the manner of Edmund Burke, Joseph Addison, and Richard Steele—an expression of his concern that early modern politics had begun to unleash dangerously disruptive passions, passions that perhaps reached their fullest expression in the French Revolution of 1789.[37] Accordingly, just as Hume recognizes the intrinsic potential for self-subversion, disruption, and deconstruction inherent in thinking and feeling, so too does the plan of government he articulates in 1752 his essay, "On the Idea of a Perfect Commonwealth," acknowledge the perennial threat of faction and discord and disruptive enthusiasm in the body politic.

But acknowledging contingency, discord, incongruity, subversion, and dissonance is not the same as valorizing them. Just as he acknowledges contingency and fragility in the laws of nature, Hume accepts the actual and potential existence of faction in the public order. He aspires, however, to structure government so that faction, while never completely eradicated, can gain little hold and the conventions of justice become natural. (In this way Hume is very much like Madison and very much a modernist.[38])

---

[37] Addison and Steele were editors and authors for the great popular periodicals, *The Tatler* and *The Spectator*, and much of the animus of their work was to combat the pernicious effects on character and on society of faction. Burke was author of trenchant criticisms of the excesses of the French Revolution, many of which were based on his defense of the value of tradition; *Reflections on the Revolution in France* (1790).

[38] Alexander Hamilton, James Madison, John Jay, *The Federalist Papers*, edited and introduced by Clinton Rossiter (New York: NAL Penguin, 1961), 526–27. Charles A. Beard, *An Economic Interpretation of the Constitution of the United States* (New York: Collier Macmillan, 1965) 15. Douglass G. Adair tests whether Madison's tenth Federalist Paper was influenced by Hume; see Adair, "The Tenth Federalist Revisited," *The William and Mary Quarterly*, 3rd series 8 (1951); "'That Politics May be Reduced to a Science': David Hume,

Hume is a skeptic, but not a nihilist. He is an antifoundationalist who nevertheless longs for the stability and order moderns look to in foundations. Hume supports modern aspirations to human liberty while discrediting social contract theory and global claims about human rights, justice and equality—the very philosophical doctrines the Whiggish used to advance the cause of liberty.[39] Similarly, while Hume remains skeptical about the possibility of establishing univocal and lasting truths, he does not rule out the possibility of doing so, and he endorses the new institutions of scientific rationality. While Hume acknowledges the pervasive and permanent character of disruption and fracture in both scientific and political theory, he remains unwilling to valorize or to promote such subversives. In this mitigated modernism Hume is a fascinating and perhaps conflicted figure, a figure who trenchantly criticizes modern modes of theory but who neverthe-less advances modern aspirations in a more moderate, skeptical, and limited form. Hume strains against the philosophical underpinnings of modernity, but he is not prepared to abandon the modern project in toto.

VI. Common Life vs. Transcendence

Some sense of this conflicted modern can be made by understand-ing how both Hume's naturalism and his skepticism are elements of a larger philosophical category—"common life." Far from simply a return to backgammon, billiards, and sociability, the reassertion of natural belief at the end of *Treatise* Part IV marks nothing more (and nothing less) for Hume than returning to "common life" as

James Madison, and the Tenth *Federalist*," *The Huntington Library Quarterly* 20 (1957), 343–60; "The Intellectual Origins of Jeffersonian Democracy: Republicanism, the Class Struggle and the Virtuous Farmer" (Ph.D. dissertation, Yale University, 1943). Hume, *Essays Moral, Political, and Literary*, 512–29. This view has recently come under renewed scrutiny.
[39] On Hume and liberty see Donald W. Livingston, "Hume's Historical Conception of Liberty," in *Liberty in Hume's History of England*, edited by Nicholas Capaldi and Donald W. Livingston (Dordrecht: Kluwer Academic Publishers, 1990), 105–54.

the natural medium and proper center of philosophical thought.[40] In advancing this claim, Hume positions himself against the philosophical enterprise which, at least since the pre-Socratics, has aspired to transcend common life and opinion. The philosophical idiom that Hume advocates is one which instead aims for the more limited objective of "methodizing" and "correcting" the beliefs, customs, habits, traditions, feelings, and practices of common life, of acknowledging and deepening how very embedded we are in them.[41] Hence we find in Hume, as we do in Wittgenstein, a narrative of alienation and return, of pathology

[40] In this I follow Donald W. Livingston. According to Livingston's reading, Book One of the *Treatise* describes the "natural history of philosophy"—that is, its beginnings in popular belief, its subsequent aspirations to autonomous foundational reasoning, its frustration in skepticism, and, finally, finding "illumination" regarding the ultimacy of common life. Donald W. Livingston, *Hume's Philosophy of Common Life* (Chicago: University of Chicago Press, 1984), 28. In Livingston's words: "The point of the Pyrrhonian arguments is not to show...that, for Hume, a person is forced to believe 'whatever nature leads him to believe, no more and no less.' Their function is, rather, to illuminate the nature of true philosophy. We may speak, then, in Hume of a 'Pyrrhonian illumination' which reveals something of the nature and limits of philosophical inquiry....From the perspective of Pyrrhonian doubt, the philosopher can see for the *first* time the magnificent, philosophically unreflective order of common life in opposition to whatever order is constituted by autonomous philosophical reflection and with an authority all its own to command belief and judgment."

For Livingston, Hume's "illumination" and return to common life mark his way of overcoming skepticism. In this I depart from Livingston, my own position being that Hume's embrace of the natural, historical, and customary world of common life is, in fact, continuous with his skepticism and exhibits Hume's willingness to continue philosophizing within its bounds. Hume writes: "In all the incidents of life we ought still to preserve our scepticism....Nay if we are philosophers, it ought only to be upon sceptical principles, and from an inclination, which we feel to the employing ourselves after that manner" (*T* 270). In short, Livingston makes a mistake analogous to Popkin's. As Popkin fails to understand how Hume's naturalism is underwritten by skepticism, so Livingston fails to understand how custom, tradition, and common life are also in Hume (unlike Burke) presented in a skeptical manner.

[41] I have in mind gestures such as Anaximander's speculations about an *apeiron* and Parmenides's reflections on the "Way of Truth." In this, Descartes follows. Hume, by contrast, represents a radical departure from this model. Hume seems to have been aware of the radicality of his thought in writing of the *Treatise* that its principles are "so remote from the vulgar Sentiments on this Subject, that were they to take place, they wou'd produce almost a total Alteration in Philosophy." This remark is nonsensical, or at least a gross exaggeration, if one simply sees Hume as a modification of Berkeley, Locke, Shaftesbury, Butler, and Malebranche.

and therapy. Hume's remark in the first *Enquiry* (1748) is characteristic:

> Happy if she [philosophy] be thence sensible of her temerity, when she pries into...sublime mysteries; and leaving a scene so full of obscurities and perplexities, return, with suitable modesty, to her true and proper province, the examination of common life. (*EHU* 81 [103])[42]

In this too Hume follows Sextus who, when discussing the skeptical four-fold and elsewhere, speaks in a strikingly similar way of *ho bios ho koinos* or the common life.[43] Hume encourages philosophy to tether itself to the realm of common life, to resist metaphysical speculation and the temptation to pursue autonomous and foundational forms of reasoning. Hume's skepticism discloses philosophy's vanity in claiming to have achieved *a priori*, ahistorical, or metaphysical knowledge. Instead, for Hume human thought must find its place within the "gross earthy mixture" (*T* 272) of common life. To the extent that, for Hume, our knowledge of others and the world is limited, we can find no a priori or demonstrative certainty. Our epistemic and moral abilities seem finite and dubious. The limits as well as the expanse of human possibility are discovered not through revelation or through

---

[42] In this I also depart from Peter F. Strawson who in *Skepticism and Naturalism: Some Varieties*, The Woodbridge Lectures 1983 (New York: Columbia University Press, 1985), recognizes significant similarities between Hume and Wittgenstein. See Section 1.4, "Hume and Wittgenstein," pp. 14–21. Strawson regards Hume's appeal to nature as a "refuge" (12) from skepticism. His reading is, therefore, consistent with Livingston's. Again, I maintain that Hume's naturalism is an extension of Hume's skepticism. One reason for Strawson's error is his oversimplifying the content of Hume's notion of "common life," reducing it to (1) belief in the existence of the body and (2) the reliability of induction (18).

[43] Sextus Empiricus's most significant usage in the *Outlines* of *ho bios ho koinos* or "common life" appears at *PH* I:237. Hume uses the phrase on many (in fact, 79) occasions, as he does its cognates. Consider, for example, Hume's important methodological claim in the *Treatise*: "We must therefore glean up our experiments in this science from a cautious observation of human life, and take them as they appear in the common course of the world" (*T* xix). Consider as well: "philosophical decisions are nothing but the reflections of common life, methodized and corrected" (*EHU* 130 [162]).

transcendental reasoning but, rather, in the same manner that our fundamental conventions are acknowledged—i.e. through nothing more or less than ordinary, common life and experience.

Hume seems conflicted because he acknowledges that we find in the experience of our lives *both* difference as well as communion, expansive possibility as well as finitude. And even our findings about such things, where we draw the line, where we cede our ground and where we claim it, remain, it seems, somehow contingent. At any point, with the advent of any new situation, the two rowers in Hume's story may find themselves out of synch, at cross purposes, inscrutable to one another. Ours is a fragile fate, despite the efforts of philosophy—and in part *because* of the efforts of philosophy—to endow it with greater security and stability, the sort of stability to which Plato and Descartes aspired in their attempts to flee the instability of appearance and opinion.[44] Hume resists the temptation to put an end to, or at least to gain respite from, the complexities and frustrations of common life.[45]   In doing so he embraces skepticism, and he exhibits a kind of *amor fati*, a courageous engagement with the human fate, even where it may disappoint us. Skepticism, Hume acknowledges, "is a malady, which can never be radically cur'd"—though many have vainly attempted to avoid or transcend it (*T* 218).

The common world in which we find a natural capacity for conventions and convention-making, as well as the possibility of their failure, is what for Hume both grounds and contains true philosophy; it is where true philosophy both begins and ends and finds a place in which to flourish. Common life is the natural

---

[44] Plato is, of course, bound up in the aspiration for transcending the human in his desire for philosophical thought to enable us to move from unstable opinion (*doxa*) to unshakeable knowledge (*episteme*), from image to reality, and from sensation to intellectual grasp of transcendent forms.

[45] By contrast, Wittgenstein writes: "The real discovery is the one that makes me capable of stopping doing philosophy when I want to.—The one that gives philosophy peace [*Ruhe*], so that it is no longer tormented [*gespeitscht*] by questions which bring *itself* in question" (*PI* #133).

element of philosophy.[46] It is, however, full of vulnerabilities and fragilities, even the sciences of common life. As philosophers we remain subjects of the empire of imagination, but we philosophers are also people who live and may live well as participants in common life.

We have seen, then, that contingency and fragility are important features of Humean philosophy. While commentators have recognized the manner in which contingency is related to custom and imagination, they have generally failed to see how contingency alloys even Hume's theorizing about nature and the natural. This failure is rooted in another, namely the failure to understand the manner in which Hume philosophizes. Hume's continuing to engage in philosophical reflection and theorizing does not mark his having rejected or overcome skepticism. Rather Hume's theorizing is itself skeptical and implies no "dogmatical spirit." Nature, therefore, does not for Hume defeat skepticism. Humean naturalism is developed as part of an acknowledgment he is led to through his skepticism—an acknowledgment of human finitude, our embeddedness in common life, and the way in which a reflective engagement with common life comprises the truest form of philosophy. It is in this acknowledgment that Hume's greatest philosophical achievement lies.

---

[46] Hume's effort to return philosophy to common life finds an analog in Heidegger's attempt to reachieve genuine "thinking" by returning thought to "its element." In his 1947 "Letter on Humanism," this Humean sentiment finds its expression in passages like this: "Thinking is judged by a standard that does not measure up to it. Such judgment may be compared to the procedure of trying to evaluate the nature and powers of a fish by seeing how long it can live on dry land. For a long time now, all too long, thinking has been stranded on dry land. Can then the effort to return thinking to its element be called irrationalism?" In Martin Heidegger, *Basic Writings: From* Being and Time *(1927) to* The Task of Thinking *(1964)*, edited by David Farrell Krell (New York: Harper & Row Publishers, 1977), 195.

EMPIRICISM
WITH AND WITHOUT
OBSERVATION
Experiments
secundum imaginationem in
Experimental Philosophy and
Demonstrative Science in
Seventeenth-Century Britain

James G. Buickerood
University of Missouri, St. Louis

Aussi en l'estude que je traitte de noz mœurs et mouvemens, les tesmoignages fabuleux, pourveu qu'ils soient possibles, y servent comme les vrais. Advenu ou non advenu, à Paris ou à Rome, à Jean ou à Pierre, c'est tousjours un tour de l'humaine capacité, duquel je suis utilement advisé par ce recit.
— Michel de Montaigne[1]

---

[1] Michel de Montaigne, "De la force de l'imagination," *Essais*, ed. Maurice Rat (Paris: Garnier, 1962), 1: 110.

[The Laputians] are very bad Reasoners, and
vehemently given to Opposition, unless when they
happen to be of the right Opinion, which is seldom
their case. Imagination, Fancy, and Invention, they
are wholly strangers to, nor have any Words in their
Language by which those Ideas can be expressed; the
whole compass of their Thoughts and Mind being
shut up within the two forementioned Sciences [i. e.
"Mathematicks and Musick"].
    — Jonathan Swift[2]

I: Introduction

This essay is an attempt to make a case in support of a claim that has long been assumed to be obviously false. It consists of a sketch of an argument for the position that Isaac Newton and John Locke each understood imagination to have beneficial, nontrivial cognitive roles—that is roles clearly within the range of that power's well-recognized representational and combinatory functions—helpful if not necessary to the production, illustration, and confirmation of knowledge and belief and motivation in experimental philosophy and in demonstrative science, respectively. More specifically, my thesis is (i) that Newton and Locke relied upon the use of what we now call "thought experiments," what were called in the scholastic tradition experiments *secundum imaginationem*, Newton in the acquisition and confirmation of scientific knowledge and Locke in the use to which knowledge is put by free agents; and further, (ii) that Newton and Locke conceived the formulation of these thought experiments to be dependent upon the proper function of imagination in the cognitive agent undertaking scientific inquiry and morally adjudicable action—or at the very least, that their respective

---

[2] Jonathan Swift, *Travels into Several Remote Nations of the World. In Four Parts.* By Lemuel Gulliver (London: for Benjamin Motte, 1726), 2: 27–28.

concepts and assessments of imagination do not preclude the possibility that imagination functions in such cognitively valuable ways. Part (i) of my thesis is to some degree perfectly conventional; part (ii) is controversial.

My interest here is not in the details or cogency of those thought experiments I shall discuss, nor in the details of the particular positions these men used those experiments to establish. Neither shall I attempt to examine or so much as to characterize all thought experiments of these two writers; there are a great many, especially of Locke's, that I shall ignore.[3] I am here interested only in the nature of those experiments I shall discuss as functions of imagination, and in what evidence there may be that Newton and Locke may have conceived them to be functions of imagination. Despite the arguably, or undeniably disparaging remarks Newton and especially Locke made about *some* imaginative processes, we shall see that they each took there to be cognitively indispensable, even valuable, functions of imagination pertinent to the acquisition and use of knowledge. These authors' notorious denigratory comments, such as they are, do not constitute wholesale condemnations of the faculty of imagination with respect to cognitive purposes. They must be read with an eye focused clearly on the immediate contexts and precise wording of those comments in order to come to an adequate understanding of these men's views.

II: Late Seventeenth- and Early Eighteenth-Century Conceptions of Imagination

What I believe can be identified as the usual understanding of early modern views of the imagination is the following. Consistent with

---

[3] My colleagues David Soles and Katherine Bradfield have helpfully examined some of Locke's thought experiments in "Some Remarks on Locke's Use of Thought Experiments," *Locke Studies* 1 (2001): 31–62.

the broad Aristotelian scheme that shifted little for nearly two millennia, imagination was generally conceived to be fundamentally a passive, representational power, the means by which the mind forms images of sensible objects (that is material objects and their properties) used in memory and, in some accounts, as one or another sort of necessary intermediary for the cognition of sensible objects.[4] One senses, reasons about, and is otherwise able to think of a material object by means of a sensory image of that object; one remembers that object by means of a sensory image held in memory (which on many accounts is generated by imagination). This is the sort of function to which Bacon alluded in his famous characterization of imagination as a messenger between reason and sense.[5] Such images were sometimes conceived to resemble their objects, sometimes precisely by way of the image's features bearing one-to-one correspondence with the features of its object. Imagination was thus conceived to be closely related to body; its objects were generally understood to be material objects, and that power itself was in some cases construed as a bodily organ, or manifestation of an organ. However that was understood in detail, imagination was almost universally conceived in terms of mental manifestations or effects of bodily phenomena. This view is accurately relayed in Ephraim Chambers' *Cyclopædia* account of imagination as a "Power or Faculty of the Soul, by which it conceives, and forms Ideas of Things, by means of certain Traces and Impressions that had been before made in the Fibres of the Brain by Sensation."[6] Imagination was not normally considered to have the capacity to create images *de novo*. What James Engell and other scholars have dubbed the "creative imagination" awaited the

---

[4] The general scheme of this concept of imagination owes to Aristotle, especially his *De Anima*, III. For not atypical early modern accounts see Thomas Hobbes, *Leviathan*, ed. Richard Tuck (Cambridge: Cambridge University Press, 1996), 15–19; and Nicolas Malebranche, *Recherche de la vérité*, ed. Geneviève Rodis-Lewis (Paris: Vrin, 1962), 1: 190–378.
[5] Francis Bacon, *The Advancement of Learning*, II, in James Spedding, Robert Leslie Ellis and Douglas Denon Heath, eds., *The Works of Francis Bacon* (London: Longman and Company, 1859), 3: 382.
[6] Ephraim Chambers, *Cyclopædia: or, an Universal Dictionary of Arts and Sciences* (London: J. and J. Knapton, 1728), 2: 375.

late eighteenth-century accounts of Tetens and Kant, arguing that imagination is a synthetic power necessary for the acquisition of knowledge.[7]

The power of imagination was also commonly thought to provide for the combination and decomposition of such images in ways not found in nature; we can form images of centaurs and flying hamsters, adding representations of features of one sort of entity to representations of the body of another sort; we can form images of eyebrows, absenting the remainder of the face of which they were originally sensed as components. Some writers found in this function of imagination profound dangers:[8] since material objects and their properties are perceived by means of images, and since images can be framed without any immediately occurrent sensory perception, the mind may mistake the latter for the former, and thus one may be induced to believe that one is sensing something that one in fact is not sensing. An often identified safeguard against this danger is an assumed regulatory function of higher cognitive faculties, or the principle that sense images are more lively and vivid than imagination images, the latter being a function of what, following the lead of Aristotle in *De Anima*, Hobbes called *"decaying sense."*[9] However, the former safeguard could be made fully intelligible only with a detailed account of the interrelative functions of the cognitive faculties and their relations

[7] Johann Nicolaus Tetens, *Philosophische Versuche über die menschliche Natur und ihre Entwicklung* (Leipzig:M. G. Weidmanns, Erben ad Reich, 1776–77); Immanuel Kant, *The Critique of Pure Reason*, especially second edition; idem, *Critique of Judgment*; James Engell, *The Creative Imagination: Enlightenment to Romanticism* (Cambridge: Harvard University Press, 1981), 119–39; Rudolph A. Makkreel, *Imagination and Interpretation in Kant: The Hermeneutical Import of the Critique of Judgment* (Chicago: University of Chicago Press, 1990).

[8] See, for example, Joseph Glanvill, *The Vanity of Dogmatizing* (London: Henry Eversden, 1661), 67, 93, 97, and 200.

[9] Aristotle, *De Anima*, III.3 (429ᵃ); Hobbes, *Leviathan*, 15. I ignore the fact that there has recently been debate over whether or not imagination (*phantasia*) was considered by Aristotle himself to be a faculty; there is no doubt that most medievals and early moderns took him to have maintained that it was. For this recent debate see Michael V. Wedin, *Mind and Imagination in Aristotle* (New Haven: Yale University Press, 1988); Malcolm Schofield, "Aristotle on the Imagination," in Martha C. Nussbaum and Amélie Oksenberg Rorty, eds., *Essays on Aristotle's 'De Anima'* (Oxford: Clarendon Press, 1992), 249–77; and Dorothea Frede, "The Cognitive Role of *Phantasia* in Aristotle," in *Essays on Aristotle's 'De Anima'*, 279–95.

with the cognitive agent himself. A satisfactory, comprehensive account of this sort was not really forthcoming in the absence of some appeal to some concept of consciousness. And the latter safeguard by itself provided in principle no real relief, for it was also recognized by at least some that imagination images could be at least as strong and vivid as sense images.[10] Consequently imagination was commonly conceived to be responsible for some sorts of non-veridical perception and cognitive errors derived from such perception.

Imagination was closely associated with the passions, at the least because humans were understood to naturally have affective responses to representations. Sometimes because it was identified as part of the body, in all cases because it was identified as a means by which the mind represents material objects and properties, imagination was considered particularly close to, influenced by and influential upon, the passions, and was understood to be capable of functioning actively on its own. The "enormous strength of Imagination" enabled it to all too readily escape the confines of reason and sense and, independently of these powers, profoundly influence the mind and the body.[11] Consequently imagination was closely associated with cognitively degraded phenomena such as madness and enthusiasm; with perceptual error and cognitive error derived from such non-veridical perception; and with the gestation and birth of deformed and monstrous infants. Part of the rationale for the latter connection with imagination, at least as old as Aristotle, was that such

---

[10] On this see, for example, possibly, Hobbes, *Leviathan*, 15–19; see *Elements of Philosophy*, in *The English Works of Thomas Hobbes*, ed. Sir William Molesworth (London: J. Bohn, 1839), 1: 405ff.; and certainly, John Locke's journal entry of 22 January 1678. Here, imagination may add "what Ideas it pleases to its owne workmanship," and "makes originals of its owne which are usually very bright and cleare in the minde and sometimes to that degree that they make impressions as strong and as sensible as those Ideas which come immediately by the senses from externall objects soe that the minde takes one for tother its own imaginations for realitys." Kenneth Dewhurst, *John Locke (1632–1704) Physician and Philosopher. A Medical Biography* (London: Wellcome Institute, 1963), 101. See also Newton's notes on Hobbes, quoted below.

[11] Henry More, *Enthusiasmus Triumphatus*, in *A Collection of Several Philosophical Writings of Dr. Henry More* (London, 1662), 4.

infants bear "false resemblances" to another species;[12] the explanation for the former stems from imagination's propinquity to the
passions, in particular to its propensity to influence and to be
influenced by them.

It is largely due to the culpability of imagination in perceptual
and cognitive error and the delusions of madness and enthusiasm
that that faculty has been understood to have been given no
important role in late seventeenth- and early eighteenth-century
philosophy.

III: Argumentum per Experimentum Secundum Imaginationem

But imagination was conceived to have important, cognitively
constructive roles in early modern philosophy. Here I shall focus
on some of those roles in natural philosophy and in demonstrative
science. And some of these roles are congruent with the tradition
of the imaginative construction of experience used in one or
another version of empiricism for centuries. Imagination's
representational function as the means by which one may seem to
sense a material object or set of circumstances and phenomena in
the natural world had not infrequently been invoked by natural
philosophers in the middle ages under the principle of *argumentum
per experimentum secundum imaginationem*. Often this consisted of
imagining or articulating (or both) unrealized physical phenomena
and relationships among material objects and properties. Medievals considered the most powerful instrument of research in
natural philosophy to be Aristotelian reason. The natural philosopher was to become acquainted with how bodies that undergo
change appear to be, and then to determine what makes them

---

[12] Aristotle, *Generation of Animals*, IV, iii (769^b–770^b). On early modern discussions of
imagination and monstrosity see Marie-Hélène Huet, *Monstrous Imagination* (Cambridge:
Harvard University Press, 1993); and Dennis Todd, *Imagining Monsters: Miscreations of the
Self in Eighteenth-Century England* (Chicago: University of Chicago Press, 1995).

appear so. The former was accomplished by empirical means, the latter through a priori reasoning using data obtained by those empirical means. Little to no controlled experiment or systematic observation contributed to this procedure. Instead, this version of Aristotelian empiricism began with a modicum of observation and experience as a basis for generalization.[13]

This commitment to observation and experience was decidedly not a commitment to experimental *practice*, so much as to experimental *arguments*. For, broadly speaking, the medievals' antecedent commitment to an Aristotelian conception of scientific knowledge precluded any movement toward a practice of controlled experiment. The nature of an object is to be determined through apprehension of the behavior of that object in its natural state, and artificial constraints on that behavior such as those imposed by controlled experiment would simply interfere with that behavior and occlude the nature of the object rather than reveal it. Thus Aristotelian *experiences* are articulations of how things occur in nature, not of something that occurs at a particular time under specified circumstances.

Medieval experimental arguments were largely based not upon personal experience (though there was more use of that in medieval natural philosophy than one might assume), but upon experience and observation reported in earlier classical, Hellenistic, and Islamic works as well as upon Western medievals' *imagined* experience and observation. We can see this in the experiments according to imagination offered by natural philosophers such as Jean Buridan, Albert of Saxony, and Nicole Oresme.[14] Buridan's thought experiment to prove the natural impossibility of a vacuum provides a relatively uncomplicated, straightforward example of a

---

[13] On medieval experimentation, see A. C. Crombie, *Styles of Scientific Thinking in the European Tradition: The History of Argument and Explanation Especially in the Mathematical and Biomedical Sciences and Arts* (London: Duckworth, 1994), 1: 313–423.

[14] See, for example, Buridan, *Questions on the Physics*, bk. 4, qu. 7, in Edward Grant, *A Source Book in Medieval Science* (Cambridge: Harvard University Press, 1974), 326; Albert, *Questiones in octo libros Physicorum*, bk. 8, qu. 12; Oresme, *Tractatus de configurationibus qualitatum et motuum*, part 2, ch. 39, in *Nicole Oresme and the Medieval Geometry of Qualities and Motions*, ed. and tr. Marshall Clagett (Madison: University of Wisconsin Press, 1968).

kind of experiment that was actually carried out centuries later by Otto von Guericke in support of the denial of Buridan's conclusion:

> If all the holes of a bellows were perfectly stopped up so that no air could enter, we could never separate their surfaces. Not even twenty horses could do it if ten were to pull on one side and ten on the other; they would never separate the surfaces of the bellows unless something were forced or pierced through and another body could come between their surfaces.[15]

Even though Guericke established the contrary of Buridan's conclusion with the aid of a similar, though actual, experiment, it is not the case that Buridan's conclusion fails simply because he attempted to establish it with a *thought*, rather than with an *actual* experiment.[16] Instead, the trouble with Buridan's experiment lies in his inference from the supposed phenomena, not with the details of the phenomena as they were imagined. Needless to say, this is a potential danger for actual experiments no less than for thought experiments.

Such thought experiments in which I am interested in the following extend from experiments that are imagined under conditions that are contrary to nature, to those that are contrived from circumstances that are credible and have some ground in observation but were nonetheless imagined for purposes of inquiry. Most *could* conceivably occur, and so functioned as if they are bona fide observations.[17] They tend to fall into one or another

---

[15] Buridan, *Questions on the Physics*, bk. 4, qu. 7.

[16] Otto von Guericke, *Experimenta nova (ut vocantur) Magdeburgica de vacuo spatio* (Amsterdam: Jansson and Wæsberge, 1672), pt. 3, "De propriis experimentis."

[17] For discussions of thought experiments in the middle ages, see Crombie, *Styles of Scientific Thinking*; Henri Hugonnard-Roche, "L'hypothétique et la nature dans la physique parisienne du xive siècle," in Stefano Caroti and Pierre Souffrin, eds., *La nouvelle physique du xive siècle* (Florence: Olschki, 1997), especially 175–77; Amos Funkenstein, *Theology and the Scientific Imagination from the Middle Ages to the Seventeenth Century* (Princeton: Princeton University Press, 1986), 152–78; John E. Murdoch, "Philosophy and the Enterprise of Science in the

of two classes: experiments *secundum imaginationem* may include descriptions of phenomena that are *in principle* unrealized, or descriptions of phenomena that are simply *contingently* unrealized. Each sort was equally considered a function of the cognitive agent's imaginative capacity; each sort was equally considered to reveal something about the nature of objects or at least about our concepts of their nature. Neither sort were generally conceived to have been simply exercises in conceptual analysis.

Such thought experiments were used in the early modern era, for example by Galileo, without overt reference to any doctrine of imagination, or indeed to any systematic philosophy of science, on which they were supposed to be based.[18] But in the middle ages there does seem to have been an elaborate, stylized study of metamethodological principles of inquiry into nature which included rules for the use of experiments *secundum imaginationem*.[19] These studies are evidently contained in that still-mysterious class of texts, the *obligationes*.[20] Though the medievals largely eschewed discussion of the mentalistic features of such experiments in favor of elaborating these experiments as sets of sentences exhibiting analyzable logical relations, they were on the whole doubtless

Later Middle Ages," in Y. Elkana, ed., *The Interaction Between Science and Philosophy* (Atlantic Highlands, NJ: Humanities Press, 1974), 51–74; idem, "The Analytic Character of Late Medieval Learning: Natural Philosophy Without Nature," in Lawrence D. Roberts, ed., *Approaches to Nature in the Middle Ages* (Binghamton, NY: Center for Medieval and Early Renaissance Studies, 1982), 171–213; Peter King, "Mediæval Thought-Experiments: The Metamethodology of Mediæval Science," in Tamara Horowitz and Gerald J. Massey, eds., *Thought Experiments in Science and Philosophy* (Savage: Rowman and Littlefield for the Center for Philosophy of Science, 1991), 43–64; and Edward Grant, *God and Reason in the Middle Ages* (Cambridge: Cambridge University Press, 2001), 168–82.

[18] For example, *Galileo Galilei 'On Motion' and 'On Mechanics', Comprising 'De Motu' (ca. 1590) and 'Le Meccaniche' (ca. 1600)*, trans. I. E. Drabkin and Stillman Drake (Madison: University of Wisconsin Press, 1960), 97–98.

[19] See especially King, "Mediæval Thought-Experiments," 50–55; and two essays in Norman Kretzmann, Anthony Kenny and Jan Pinborg, eds., *The Cambridge History of Later Medieval Philosophy* (Cambridge: Cambridge University Press, 1982): Eleonore Stump, "Obligations: From the Beginning to the Early Fourteenth Century," and Paul Vincent Spade, "Developments in the Fourteenth Century", 315–41.

[20] For example, Romuald Green, *The Logical Treatise 'De Obligationibus': An Introduction with Critical Texts of William of Sherwood and Walter Burley* (St. Bonaventure: Franciscan Institute, 1982).

cognizant of some mentalistic basis for the practice, as the common characterization of such experiments as "experiments according to imagination" itself suggests.

Galileo notoriously asserted that an experiment poses a question to nature. If we broaden this view in such a way as to accommodate early moderns' use and understanding of experiment, we may define an experiment as a procedure for raising or resolving a question about relations between specified variables by deliberately altering the value of one of them and articulating the response to this alteration by the other variable(s). And a thought experiment is just such an experiment that attempts to achieve this goal without *actually* executing that process in nature.[21] Specifically, instead of sensorily or instrumentally witnessing the occurrent phenomena of the experiment, the thought experimenter imagines witnessing the phenomena of the experiment. More simply, he imagines what constitutes the experiment. The only cognitive power available in the early moderns' (and in the medievals') conception of mind by which phenomena such as these could possibly be contrived is the imagination, which is thus capable of providing an empiricism without observation.

IV: Inadequacies of Our Usual Understanding of Early Modern Concepts of Imagination

Our conventional understanding of early modern views and uses of imagination can accommodate contemporary use of thought

---

[21] Philosophical studies of thought experiments are available in Roy A. Sorensen, *Thought Experiments* (N. Y.: Oxford University Press, 1992); James Robert Brown, *The Laboratory of the Mind: Thought Experiments in the Natural Sciences* (London: Routledge, 1991); and Tamara Horowitz and Gerald J. Massey, eds., *Thought Experiments in Science and Philosophy*. There is disagreement both on the analysis of thought experiments and on their roles in conceptual and empirical studies. Sorensen in particular argues forcefully for the position adopted here, that thought experiments are bona fide experiments, and both he and Brown take the contribution of thought experiments to the scientific study of nature to be very important.

experiments, though it is not sufficiently recognized that they owe to that cognitive power. Our conventional understanding cannot, however, be altogether reconciled with the evidence of some important texts which exhibit the use of *other* quite important functions of imagination. Remarkable examples of a concept of imagination that transgress the boundaries of the commonplace interpretation may be found in the work of René Descartes. Descartes not only wrote in detail of his physical hypotheses and of his elaborate thought experiments, both derived, on his own account, from the faculty of imagination, but his own conception of *ideas* was articulated as a species of *images*. Also, in at least two of his works, Descartes attributed to imagination a synthetic function more akin to Kant's notion of imagination than the stereotypical writer of the early modern period who took imagination to be exclusively the representation of material objects by means of resemblances which are images.[22] In one of these latter works, the *Regulæ ad directionem ingenii* (*Rules for the Direction of the Mind*), Descartes articulated one function of imagination not to complement or to be subordinated to the operation of reason, but to obtain *within* the operation of reason. In the seventh rule, imagination is required in order to effect an intuitive perception of a deductive series. An act of deducing often requires the apprehension of a lengthy series of conceptual transitions from ground to consequent. In such cases, the cognitive agent finds it difficult to recall ("non facilè recordemur") the entire series of conceptual connections from, for example, premisses to conclusions ("totius itineris"). Therefore, Descartes proposed, there "must be a continuous movement of thought" to rectify this weakness of memory and make possible deductive inference from known grounds ("ideoque memoriæ infirmitati continuo quodam cogitationis motu succurrendum esse dicimus"). Should one, for instance, *intuit* the relation between successive elements of a series from *A* to *E*,

---

[22] Descartes' explicit use of imagined hypotheses—some even false—is nearly ubiquitous, but see particularly *Traité de l'homme*; for ideas as images, "tanquam rerum imagines sunt," see meditation 3; for his use of imagination as a synthetic power, see *Regulæ ad directionem ingenii*, rule 7, and *Compendium musicæ*, as cited below.

intuiting the relation between A and B, then distinctly intuiting the relation between B and C, intuiting that between C and D, and once further between D and E, these separate intuitions, even in series, do not *constitute* one's apprehending the relation between A and E, nor do they *entail* one's apprehending this relation. Only having all these elements present to mind simultaneously and in their proper order can yield a precise knowledge ("intelligere præcisè") of the relation between A and E. Memory itself, said Descartes, cannot reliably supply all this material to mind. To remedy that deficiency, the imagination, he said, must provide a *composite intuition*:

> To remedy this [weakness of memory], I would run...[the relations] over from time to time, keeping the imagination moving continuously in such a way that while it intuitively perceives each fact it simultaneously passes on to the next; and this I do until I learn to pass from the first to the last so swiftly that no role in this process is left to memory, and I intuit the whole at once. This method relieves memory, diminishes the sluggishness of our thinking, and enlarges our mental capacity.[23]

The imagination is to be kept in continuous motion so that one may *thereby intuitively perceive* all relata and relations simultaneously ("singula intuentis simul"), instead of having a series of discrete intuitions of the relations and relata with no connections between these intuitions present to mind. The imagination yields the intuitive perception of the entire series as a whole ("rem totam

---

[23] "Quamobrem illas continuo quodam imaginationis motu singula intuentis simul & ad alia transeuntis aliquoties percurram, donec à primâ ad vltimam tam celeriter transire didicerim, vt ferè nullas memoriæ partes relinquendo, rem totam simul videar intueri; hoc enim pacto, dum memoriæ subvenitur, ingenij etiam tarditas emendatur, ejusque capacitas quâdam ratione extenditur." Descartes, *Regulæ directionem ingenii*, in Charles Adam and Paul Tannery, eds., *Œuvres de Descartes*. Nouvelle présentation (Paris: Vrin, 1974–1982), X, 388. All following references to this edition will conform to the conventional form of "AT," followed by a Roman numeral indicating volume number, succeeded by an Arabic numeral denoting the page number, rendering this instance: AT X, 388.

simul videar intueri"), making it possible for one to know not only the relations between, for example, *A* and *B*, but all consequent conceptual connections, including that between *A* and *E*. The limitations not merely of memory, but of reason itself—specifically of intuition—are rectified by the synthetic function of imagination. Deductive knowledge is therefore possible only on the condition that imagination provides the synthesis of discrete objects of thought and their relations.

Descartes appealed to such a synthetic function of imagination again in his *Compendium musicæ* of 1618, the young philosopher's inquiry into proportions or ratios. Here, imagination again synthesizes a series of disparate elements otherwise only perceptible individually, and not as a whole.[24] It is the means by which one is able to hear a melody, to perceive the complex unity of sounds as a unity rather than as a mere congeries of disparate tones.

The example of Descartes ought to make us suspicious of the adequacy of the usual understanding of imagination in the early modern era as largely representational and combinatory, used exclusively to represent material objects and their properties. The preponderance of thought experiments (of so many kinds) in the philosophical and scientific literature of the seventeenth and eighteenth centuries ought to make us suspicious of conventional interpretations of early moderns' own denigrations of that power. The precedent of the use and doctrine of experiments according to the imagination in the high middle ages can only deepen such suspicions. And the particularities of the texts of many of those early modern denigrations themselves ought to make us circumspect and curious about their authors' considered views of imagination as a whole.

---

[24] *Compendium musicæ*, part 3: "De Numero vel tempore in sonis observando" ("Of the Perception of Number or Time in Sounds"), AT X, 93–94.

V: Newton

L'Ouvrage de Monsieur Newton est une Mecanique
la plus parfaite qu'on puisse imaginer.
— Journal des sçavans, *1688*[25]

It did not take long for readers of Isaac Newton's *Principia Mathematica* to interpret its author's scattered methodological remarks in that book to include a dismissal of imaginative functions from the successful pursuit of experimental philosophy, or that study of natural philosophy that inquires into the causes of things no further than they can be "proved by Experiments."[26] It is a reading that has currency to this day. This sort of interpretation seems to have been produced from varied motivations, so far as motivations can be identified at all. In some cases, the claim or implication that Newton dismissed imaginative products was favorably contrasted with Descartes' profligate construction of "a purely imaginary physics, a philosophical romance," as Huygens and Voltaire had it.[27] Sometimes this interpretative trope focused on Newton's supposed refusal to "feign"—read as a refusal to *imagine*—hypotheses, in contrast with Descartes' shameless proliferation of explicitly imagined hypotheses and suppositions. Sometimes it has more straightforwardly been supposed to have been of a piece with what has been incorrectly understood as Newtonian empiricism to largely consist of the prosecution of a hypothetico-deductive method. As Alexandre Koyré noted years ago, the contrived contest between Descartes and Newton has

transformed both of them into symbolic figures; the one, Newton, embodying the ideal of modern, progressive,

---

[25] *Journal des sçavans* 16 (1688): 237: "The work of Mr Newton is a mechanics, the most perfect one could imagine."

[26] [Isaac Newton], "An Account of the Book entituled *Commercium Epistolicum Collinii et aliorum, De Analysi promota*," in *Philosophical Transactions* 29 (1715): 222.

[27] See Alexandre Koyré, *Newtonian Studies* (Chicago: University of Chicago Press, 1965), 62 and 63.

and successful science, conscious of its limitations and firmly based upon experimental and experimental-observational data which it subjected to precise mathematical treatment; the other, Descartes, symbol of a belated, reactionary—and fallacious—attempt to subject science to metaphysics, disregarding experience, precision, and measurement, and replacing them by fantastic, unproved, and unprovable hypotheses about the structure and behavior of matter. Or, even more simply, the one, Newton, representing the truth, and the other, Descartes, a subjective error.[28]

This is a common conception of Newton's method even absenting any express contrast with Descartes. Characterizing Newtonian method on the basis of Newton's methodological remarks rather than on his actual procedure in the *Principia*, Colin Turbayne voiced many scholars' understanding of the method: "The main features of Newton's method...are: The rejection of hypotheses, the stress upon induction, the working sequence (induction precedes deduction), and the inclusion of metaphysical arguments in physics." The defining characteristic of Newtonian method is supposed to be its deductive procedure.[29]

This reflects serious misunderstanding of the method Newton followed in the *Principia*, and it occludes his use of imagination in at least two respects, only one of which I shall examine in this essay. In the first place, Newton's method was not uniformly to subject experimental and observational data to precise mathematical treatment. Rather, the Newtonian procedure in experimental philosophy was to blend imaginative reasoning with the applica-

---

[28] Koyré, *Newtonian Studies*, 55–6. On eighteenth-century uses of and confusion over Newtonian method, see, for example, Robert Schofield, "An Evolutionary Taxonomy of Eighteenth-Century Newtonianisms," in Roseann Runte, ed., *Studies in Eighteenth-Century Culture* (Madison: University of Wisconsin Press, 1978), 7: 175–92; and Buickerood, "Pursuing the Science of Man: Some Difficulties in Understanding Eighteenth-Century Maps of the Mind," *Eighteenth-Century Life* (n. s.) 19 (1995): 1–17.
[29] Colin Turbayne, *The Myth of Metaphor* (New Haven: Yale University Press, 1962), 45; see 45–49.

tion of mathematical procedures to phenomena. This consisted in beginning with a collection of imagined material objects and conditions "simpler than those of nature, which can be transferred from the world of physical nature to the domain of mathematics."[30] The result of this is not simply an imagined construct. Whereas the scholastic natural philosophers such as Buridan had imagined phenomena and contrived mathematical systems for the purposes of investigating the consequences of the conditions they had imposed with at most an eye to determining the nature of objects, Newton's imaginative constructs were contrived in order to account for the natural world as the context and subject of experience. The latter characteristically articulated his imaginative possibilities in terms of conditionals that captured real conditions.[31] Newton's foundational methodological contribution to natural philosophy derived from his facility in separating scientific inquiry into two parts: "the development of the mathematical consequences of imaginative constructs or systems and the subsequent application of the mathematically derived results to the explanation of phenomenological reality."[32] This is Newton's most ramifying use of imagination.

The second, more conventional (in his own time and in ours) use to which Newton put imagination, a use that has been unappreciated as a form of *imagining*, is exhibited in the thought experiments he constructed for specific physical and metaphysical claims. But before discussing one of those thought experiments it would be helpful to examine just how Newton himself wrote about imagination in order to dispel at least some of the grounds

---

[30] I. Bernard Cohen, *The Newtonian Revolution, With Illustrations of the Transformation of Scientific Ideas* (Cambridge: Cambridge University Press, 1980), 62.

[31] For example, in Book 1, the first of the three axioms or laws of motion: "*Every body perseveres in its state of being at rest or of moving uniformly straight forward, except insofar as it is compelled to change its state by forces impressed [upon it].*" *The 'Principia': Mathematical Principles of Natural Philosophy*, trans. I. Bernard Cohen and Anne Whitman assisted by Julia Budenz (Berkeley: University of California Press, 1999), 416.

[32] Cohen, *Newtonian Revolution*, 66 and xii. Cf. Peter Dear's interesting discussion of the historical implications of Newton's procedure in *Discipline and Experience: The Mathematical Way in the Scientific Revolution* (Chicago: University of Chicago Press, 1995), especially 210–43.

for the interpretation that he did not believe that that faculty had any constructive contribution to make to experimental philosophy.

Consider the following two passages and their implications for Newton's conception of imagination and its potential contribution to scientific inquiry. The first:

> *In experimental philosophy we are to look upon propositions collected by general induction from phenomena as accurately or very nearly true, notwithstanding any contrary hypotheses that may be imagined, till such time as other phenomena occur, by which they may either be made more accurate, or liable to exceptions.*[33]

And the second passage:

> il [Newton] dit que pour parvenir à la connoissance du vrai systeme du monde, il ne faut pas s'en rapporter à son imagination, mais consulter la nature; que des fictions, quelque ingenieuses q[u]'elles soyent, sont pourtant des fictions, au lieu que l'experience conduit à la réalité.[34]

The first passage is from the third book of Newton's *Principia*, in Andrew Motte's translation of the fourth of Newton's "Regulæ Philosophandi"; the second is from the anonymous review of the second edition of Émilie du Châtelet's French translation of the

---

[33] Newton, *The Mathematical Principles of Natural Philosophy*, translated by Andrew Motte (London: Benjamin Motte, 1729), 2: 205. The French translation of the *Principia* was evidently influenced by Motte's diction. Mme. du Châtelet had Newton writing "Je n'imagine pas d'hypothèses," rather than "Je ne feins pas d'hypothèses." *Principes mathématiques de la philosophie naturelle*, tr. Madame la Marquise du Chastelet (Paris: Desaint et Saillant, 1759). See Koyré, *Newtonian Studies*, 36.

[34] *Mémoires pour l'histoire des sciences et des beaux arts* (Trévoux: Etienne Ganeau, 1718), 470. "[Newton] says that in order to arrive at knowledge of the true system of the world, one must not put one's faith in one's imagination, but must consult nature; that fictions, however ingenious they may be, are nonetheless fictions, whereas experience leads to reality." Translation mine.

work, characterizing this fourth of Newton's rules for the study of natural philosophy.

Together with the reading of the notorious declaration, "hypotheses non fingo," the fourth rule is taken as the basis of Newton's supposedly antagonistic attitude toward imagination in experimental philosophy. His attitude toward hypotheses and just how one is to interpret the 'non fingo' dictum has, in my view, been adequately explained by Bernard Cohen's painstaking analysis.[35] Very roughly speaking, hypotheses do play legitimate roles in *natural* philosophy, but not in its subfield, *experimental* philosophy. For my purposes here, we may confidently take it to be the case that Newton was not thereby signaling either a rejection of hypotheses from natural philosophy *tout court*, nor impugning any possible constructive contribution of imagining in experimental philosophy. A further objection to such an interpretation as this anonymous reviewer's and others', and to any use of the fourth rule as evidence for it, is that Motte failed to correctly capture what Newton actually wrote in his Latin text: Newton did not say that one must not put one's faith in *imagination*, nor did he write that one must not imagine hypotheses which would invariably be at odds with empirical data. Rather, he said that the experimental philosopher must not persisting in holding hypotheses contrary to those propositions one derives from a general induction from phenomena: *"In philosophia experimentali, propositiones ex phænomenis inductionem collectæ, non obstantibus contrariis hypothesibus, pro veris aut accurate aut quamproxime haberi*

---

[35] See, for example., Gerald Holton, "The Thematic Imagination in Science," in Holton, ed., *Science and Culture: A Study of Cohesive and Disjunctive Forces* (Boston: Houghton Mifflin, 1965), 88–108; Cohen, *Franklin and Newton: An Inquiry into Speculative Newtonian Experimental Science and Franklin's Work in Electricity as an Example Thereof* (Philadelphia: American Philosophical Society, 1956), *passim*; Cohen, "The First English Version of Newton's 'Hypotheses non fingo,'" *Isis* 53 (1962), 379–88; Cohen, "Hypotheses in Newton's Philosophy," *Physis: Rivista internazionale di storia della scienza* 8 (1966), 163–84; Cohen, *Introduction to Newton's 'Principia'* (Cambridge: Harvard University Press, 1971), especially 30, 37, 241–2, and 259–60; and Cohen, *Newtonian Revolution*, especially 100–01.

debent, donec alia occurrerint phænomena, per quæ aut accuratiores reddantur aut exceptionibus obnoxiæ."[36]

An important question is just what is to be included within this category of phenomena. The category is not restricted to what is apprehensible through sensory perception. Newton provided some indication of the breadth of this category in drafts intended for, but never included in, the *Principia*. In a draft "rule 5," Newton contended that whatever is not derived from things themselves, *either by sensory perception or by the sensation of internal thoughts*, is a hypothesis; whatever is an object of the operations of any of our various cognitive faculties is a phenomenon:

> Whatever things are not derived from objects themselves, whether by the external senses or by the sensation of internal thoughts, are to be taken for hypotheses. Thus I sense that I am thinking, which could not happen unless at the same time I were to sense that I am.... And I do not take for phenomena only things which are known to us by the five external senses, but also those which we contemplate in our minds when thinking.... And those things which follow from the phenomena neither by demonstration nor by the argument of induction, I hold as hypotheses.[37]

---

[36] Newton, *Isaac Newton's Philosophiæ Naturalis Principia Mathematica. Third edition (1726) with Variant Readings*. Assembled and edited by Alexandre Koyré and I. Bernard Cohen with the Assistance of Anne Whitman (Cambridge: Harvard University Press, 1972), 2: 555. Cohen's and Whitman's new translation, *The Principia*, has now supplanted Cajori's modified version of the long-serving Motte as the standard English translation of the work. They translate this passage as: *"In experimental philosophy, propositions gathered from phenomena by induction should be considered either exactly or very nearly true notwithstanding any contrary hypotheses, until yet other phenomena make such propositions either more exact or liable to exceptions." The 'Principia'*, 796.

[37] "Pro hypothesibus habenda sunt quæcunque ex rebus ipsis vel per sensus externos, vel per sensationem cogitationum internarum non derivantur. Sentio utique quod Ego cogitem, id quod fieri nequiret nisi simul sentirem quod ego sim.... Et pro Phænomenis habeo, non solum quæ per sensus quinque externos nobis innotescunt, sed etiam quæ in mentibus nostris intuemur cogitando.... Et quæ ex Phænomenis nec demonstrando nec per argumentum Inductionis consequuntur, pro Hypothesibus habeo." University Library, Cambridge, MS. Add. 3965. Quoted in Cohen, *Introduction*, 30; cf. Koyré, *Newtonian Studies*,

Phenomena—data for the experimental philosopher—therefore include the products of imaginative thought as well as of memory and sense and reason. This is a notion to which Newton returned more than once in versions of his preface and other unpublished comments, sometimes sketching the correlations between the various sorts of phenomena, their cognitive origins, and how they map on to contemporarily recognized philosophical disciplines:

> What is taught in metaphysics, if it is derived from divine revelation, is religion; if it is derived from phenomena through the five external senses, it pertains to physics; if it is derived from knowledge of the internal actions of our mind through the sense of reflection, it is only philosophy about the human mind and its ideas as internal phenomena likewise pertain to physics. To dispute about the objects of ideas except insofar as they are phenomena is dreaming....[38]

While the assertion that knowledge of the internal actions of the mind constitutes knowledge of internal phenomena which are included within the subject-matter of physics ("ad Physicam pertinet") suggests perhaps some Hobbesian influence (which is discernible elsewhere in his writings), Newton's considered position seems to have been rather some species of dualism. That is, he did not altogether accept Hobbes's identification of thoughts with motions in the brain; such motions are instead thought to be present to the soul in the sensorium.[39] In any case, the ontological

---

272. A transcription of the original Latin text is included in D. T. Whiteside, ed., *The Mathematical Papers of Isaac Newton* (Cambridge: Cambridge University Press, 1981), 8: 442–59.

[38] University Library, Cambridge MS. Add. 3968, fol. 109. Quoted from I. Bernard Cohen, "A Guide to Newton's *Principia*," in *The Principia*, tr. Cohen et al., 54. Prior to this paragraph Newton had written—and then deleted—the claim that the "actions of the mind of which we are conscious" ("quarum conscii sumus") are phenomena.

[39] On this see J. E. McGuire and Martin Tamny, eds., *Certain Philosophical Questions: Newton's Trinity Notebook* (Cambridge: Cambridge University Press, 1983), 216–40. I work from the MS. originals, but shall cite the locations of McGuire and Tamny's transcriptions in their edition for ease of reference.

status of such phenomena themselves is beside the point of Newton's having taken them to provide the data of experimental philosophy.

Newton's conception of the power of imagination is less than limpid. He nowhere to my knowledge articulated his conception of imagination in anything like systematic detail.[40] What emerges from his early notes is a fairly conventional early modern understanding of that faculty, touching on his study of Henry More, Joseph Glanvill, Thomas Hobbes, and René Descartes in turns.[41] His comments here include critical responses to Hobbes's theory of sense and memory as motions in the brain by countering that

> Motion is never ye weake[r] for ye object being taken away for y^n dreames would not be so cleare as sence. But to men wakeing[,] things past appeare ob[scurer] then things present because ye organs being moved by other p^r[e]sent objects at ye same time[,] those phantasms are lesse predominant....
>
> ... Then we should never forget any thing. 2. Phantasmes are prædominant from ye strength of the motion causing y^m...but if ye motion causing present & past phantasms be alike strong[,] ye effects must be equall & so there would be no differences betwixt sence & phantasie. All things wch wee ever perceived would be alike in our phantasie & wee should thinke of an immense multitude of objects at once.[42]

---

[40] Though there are early manuscript notes, now published by McGuire and Tamny in *Certain Philosophical Questions*, including "Immagination. & Phantasie & invention", "Of imagination", and "Of ye soule", especially. University Library, Cambridge. MS. Add. 3996, fols. 109, 125, and 130.

[41] Specifically, More, *The Immortality of the Soul*, in *A Collection*; Glanvill, *The Vanity of Dogmatizing*; Hobbes, *Elements of Philosophy. The First Section, concerning Body* (London: Andrew Crooke, 1656); and Descartes, *Meteora*, and *Principia Philosophiæ*.

[42] University Library, Cambridge. MS. Add. 3996, fol. 85 130^r; see *Certain Philosophical Questions*, 450.

Were Hobbes's analysis correct, not only would we be unable to distinguish sensings from imaginings phenomenologically, our senses would constantly be filled with indistinguishably present and past phantasms. So it is *quantity* of motion, thought Newton, that distinguishes sense from imagination, much like quantity of motion distinguishes sounds and colors. While nonetheless maintaining such a distinction between these two powers, Newton considered sense and imagination to have "ye same operation upon ye spirits" in the optic nerve, and that "ye same motions are caused in...braines by both."[43] Thus it would appear that while we can provide a physicalist distinction between sense and imagination, no certain phenomenological distinction correlative with this is obtainable. And while at the very least Newton's probing after-image experiments showed that imagination may involuntarily—and even unbeknownst to the subject—influence the perception of the cognitive agent under suitable conditions, imagination, like memory, may be in some ways controlled by one's volition: one may deliberately imagine some object or event.

Though imagination or fantasy is very rarely *explicitly* mentioned or appealed to in the spare Latin prose of the *Principia*, Newton's Latin papers drafting analyses that were to see publication in one or another edition of that book reveal a thinker frequently and deliberately explicitly resorting to imaginative possibilities in investigating nature. In these papers, imagination provides the natural philosopher with phenomena that he could not in principle, or because of a variety of contingencies, directly sense.[44] One may imagine sensing matter in a vortex at rest or in motion (93, 125); one may imagine space empty of body (99, 132),

[43] University Library, Cambridge. MS. Add. 3996, fol. 75 125^r; see *Certain Philosophical Questions*, 442. What follows is Newton's description of after-image experiments he relayed in a letter to Locke on 30 June 1691. See *The Correspondence of John Locke*, ed. E. S. de Beer (Oxford: Clarendon Press, 1979), 4: 288–90 (letter no. 1405).

[44] All of the following references to Newton's manuscripts owe to their transcription in A. Rupert Hall and Marie Boas Hall, eds., *Unpublished Scientific Papers of Isaac Newton. A Selection from the Portsmouth Collection in the University Library, Cambridge* (Cambridge: Cambridge University Press, 1962).

and partly by this means establish the anti-Cartesian position that extension is neither substance nor accident.

The most extensive express use of imagination in these papers is in the interesting essay, "De Gravitatione et æquipondio fluidorum" ("On the Gravity and Equilibrium of Fluids") to establish the position that space extends infinitely in all directions (91–121, 121–56). We cannot imagine any limit anywhere without at the same time understanding that there is space beyond that limit, and thus that space extends infinitely. ("Non possumus enim ullibi limitem imaginari quin simul intelligamus quod ultra datur spatium.") One cannot object that this infinity is only imaginary and not real,

> for if a triangle is actually drawn, its sides are always, in fact, directed towards some common point, where both would meet if produced, and therefore there is always such an actual point where the produced sides would meet, although it may be imagined to fall outside the limits of the physical universe. And so the line traced by all these points will be real, though it extends beyond all distance.[45]

The experimental philosopher's imagination is guided and limited by mathematical reasoning, and so is a tool providing him with phenomena otherwise unavailable. We can see, then, that Newton had no principled bias against the use of imagination to contrive possibilities, ideal objects and events for mathematical articulation and subsequent application to nature. And this is, broadly, just the role of the thought experiments he offered in the *Principia*.

---

[45] *Unpublished Scientific Papers*, 134. The Latin transcription is: "Neque est quod aliquis dicat hanc imaginatione tantum et non actu infinitam esse; nam si triangulum sit actu adhibitum, ejus crura semper actu dirigentur versus aliquod commune punctum, ubi concurrerent ambo si modo producerentur, et proinde tale punctum ubi productæ concurrerent semper erit actu, etiamsi fingatur esse extra mundi corporei limites; atque adeo linea quam ea omnia puncta designant erit actualis, quamvis ultra omnem distantiam progrediatur" (101).

The two most famous thought experiments in the *Principia* are presented in the scholium attached to the first section of that work, on definitions: the rotating bucket experiment and the whirling globes experiment.[46] The role of imagination in constructing these experiments will be amply exhibited by a glance at the whirling globes experiment. Newton opens that with the admission that it is "certainly very difficult to find out the true motions of individual bodies and actually to differentiate them from apparent motions."[47] The reason for this is that those parts of

---

[46] Newton, *The "Principia,"* 412–15. There is some question as to whether the first of these is a thought experiment, or an experiment that Newton actually performed. For example, James Robert Brown, *Laboratory of the Mind,* 6–10 and 40f., assumes that the bucket experiment is a thought experiment. Roy Sorensen, *Thought Experiments,* 144–47, assumes that it is not. Cohen, in his "Guide to Newton's *Principia,*" 107, takes Newton's assertion "as experience has shown me" ("ut upse expertus sum") in the report of the bucket experiment to entail that it is not a thought experiment. But this assertion does not clearly mean that Newton has experienced all the conditions governing the experiment. He makes this parenthetical remark within his description of the rotating water inside the bucket forming itself into a concave figure—something we can see under conditions other than those delineated in the experiment as described. One might think it more likely (though this is all speculative to a degree) that had Newton actually performed the experiment, he could have far more straightforwardly indicated that fact, instead of simply saying that he'd experienced this specific phenomenon.

An additional complication in interpreting both experiments is that the literature in the history and philosophy of science has long been infected with a misunderstanding of their purpose: they have been taken to provide the grounds for Newton's argument for the existence of absolute space and the impossibility of a relativistic mechanics. See, for example, Ian Hacking, "The Identity of Indiscernibles," *The Journal of Philosophy* 72 (1975): 249–50; Ernest Nagel, *The Structure of Science: Problems in the Logic of Scientific Explanation* (New York: Harcourt, Brace and World, 1961), 209; Hans Reichenbach, *Philosophy of Space and Time,* tr. Maria Reichenbach and John Freund (New York: Dover, 1957), 213–14; and Max Jammer, *Concepts of Space* (N. Y.: Harper, 1960), 106. Two compelling counterarguments may be found in Ronald Laymon, "Newton's Bucket Experiment," *Journal of the History of Philosophy* 16 (1978): 399–413; and Robert Rynasiewicz, "By Their Properties, Causes and Effects: Newton's Scholium on Time, Space, Place and Motion," *Studies in the History and Philosophy of Science* 26 (1995): 133–53, and 295–321.

To argue against what I believe to be misinterpretations of these two experiments would take me too far afield from my purpose in this essay, and is in any case unnecessary to illustrate Newton's use of imagination to provide phenomena for experimental philosophy. I will, then, simply assert that I take it to be the case that the rotating bucket experiment and the whirling globes experiment are both thought experiments, and that neither serves to establish the existence of absolute space and the impossibility of a relativistic mechanics.

[47] Newton, *The Principia,* 414.

immovable space in which the bodies really move are not sensed—they are imperceptible. This fact of the matter notwithstanding, all hope is not lost, for we can contrive phenomena that we can subject to analysis by drawing evidence partly from apparent motions, and partly from the "forces that are the causes and effects of the true motions." For instance,

> if two balls, at a given distance from each other with a cord connecting them, were revolving about a common center of gravity, the endeavor of the balls to recede from the axis of motion could be known from the tension of the cord, and thus the quantity of circular motion could be computed. Then, if any equal forces were simultaneously impressed upon the alternate faces of the balls to increase or decrease their circular motion, the increase or decrease of the motion could be known from the increased or decreased tension of the cord, and thus, finally, it could be discovered which faces of the balls the forces would have to be impressed upon for a maximum increase in the motion, that is, which were the posterior faces, or the ones that are in the rear in a circular motion. Further, once the faces that follow and the opposite faces that precede were known, the direction of the motion would be known. In this way both the quantity and the direction of this circular motion could be found in any immense vacuum, where nothing external and sensible existed with which the balls could be compared.[48]

If we complicate this picture by postulating the existence of some distant bodies in this space that maintain given positions relative to one another (like the "fixed stars" in the heavens), we could not know from the relative change in position of these balls among these bodies whether the motion is to be attributed to the balls or to the new bodies. But, if we were to examine the cord and

---

[48] *The Principia*, 414.

recognized that its tension was the tension which the motion of the balls required, we would be justified in inferring that the motion belonged to the balls, and that the bodies were at rest. And we could determine the direction of this motion on the basis of our observations of the change in position of the balls among the distant bodies.[49]

The importance of this derivation of phenomena by imagining becomes clear when we consider Newton's purpose in laying it out. It is not, even in conjunction with the preceding rotating bucket experiment, sufficient to establish the existence of absolute space. Rather, its purpose at the close of the scholium on time,

---

[49] All these conditionals are amply evident in the Latin original, and themselves suggest that this experiment is an imaginative construct. See *Isaac Newton's Philosophiæ Naturalis*, 1: 52–3: "Quantitates relativæ non sunt igitur eæ ipsæ quantitates, quarum nomina præ se ferunt, sed sunt earum mensuræ illæ sensibiles (veræ an errantes) quibus vulgus loco quantitatum mensuratarum utitur. At si ex usu definiendæ sunt verborum significationes; per nomina illa temporis, spatii, loci & motus proprie intelligendæ erunt hæ mensuræ sensibiles; & sermo erit insolens & pure mathematicus, si quantitates mensuratæ hic intelligantur. Proinde vim inferunt sacris literis, qui voces hasce de quantitatibus mensuratis ibi interpretantur. Neque minus contaminant mathesin & philosophiam, qui quantitates veras cum ipsarum relationibus & vulgaribus mensuris confundunt.

"Motus quidem veros corporum singulorum cognoscere, & ab apparentibus actu discriminare, difficillimum est; propterea quod partes spatii illius immobilis, in quo corpora vere moventur, non incurrunt in sensus. Causa tamen non est prorsus desperata. Nam argumenta desumi possunt, partim ex motibus apparentibus qui sunt motuum verorum differentiæ, partim ex viribus quæ sunt motuum verorum causæ & effectus. Ut si globi duo, ad datum ab invicem distantiam filo intercedente connexi, revolerentur circa commune gravitatis centrum; innotesceret ex tensione fili conatus globorum recedendi ab axe motus, & inde quantitas motus circularis computari posset. Deinde si vires quælibet æquales in alternas globorum facies ad motum circularem augendum vel minuendum simul imprimerentur, innotesceret ex aucta vel diminuta fili tensione augmentum vel decrementum motus; & inde tandem inveniri possent facies globorum in quas vires imprimi deberent, ut motus maxime augeretur; id est, facies posticæ, sive quæ in motu circulari sequuntur. Cognitis autem faciebus quæ sequuntur, & faciebus oppositis quæ præcedunt, cognosceretur determinatio motus. In hunc modum inveniri posset & quantitas & determinatio motus hujus circularis in vacuo quovis immenso, ubi nihil extaret externum & sensibile quocum globi conferri possent. Si jam constituerentur in spatio illo corpora aliqua longinqua datam inter se positionem servantia, qualia sunt stellæ fixæ in regionibus cœlorum: sciri quidem non posset ex relativa globorum translatione inter corpora, utrum his an illis tribuendus esset motus. At si attenderetur ad filum, & deprehenderetur tensionem ejus illam ipsam esse quam motus globorum requireret; concludere liceret motum esse globorum, & corpora quiescere; & tum demum ex translatione globorum inter corpora, determinationem hujus motus colligere."

space, place, and motion is to exhibit a compelling case in which the experimental philosopher can obtain evidence regarding the true motion of individual bodies on the condition that the absolute space in which they move cannot be perceived. Imagination provides in this way the experimental philosopher with data grounding genuine theoretical knowledge of nature.

VI: Locke

> *Upon a stricter enquiry, I am forced to conclude, that good, the greater good, though apprehended and acknowledged to be so, does not determine the will, until our desire, raised proportionably to it, makes us uneasy in the want of it.*   — John Locke[50]

The brief concluding chapter of Locke's *An Essay concerning Human Understanding* lays out a tripartite division of the sciences that includes all "that can fall within the compass of Humane Understanding" (4.21.1). The three sciences he identified here are,

> *First*, The Nature of Things, as they are in themselves, their Relations, and their manner of Operation: Or, *Secondly*, That which Man himself ought to do, as a rational and voluntary Agent, for the Attainment of any End, especially Happiness: Or, *Thirdly*, The ways and means, whereby the Knowledge of both the one and the other of these, are attained and communicated. (4.21.1)

Natural philosophy, ethics, and logic are the three sciences possible for humankind. Ethics, or Πρακτική, we must remember, was in

---

[50] Locke, *An Essay concerning Human Understanding*, ed. P. H. Nidditch (Oxford: Clarendon Press, 1975), 2.21.35. All further citations and references to the *Essay* owe to this edition, and shall follow the convention of citing book, chapter, and section numbers in sequence, the latter two numerals separated from the previous numerals by points. Thus, '2.21.35' refers to book 2, chapter 21, section 35.

Locke's view a *"demonstrative science,"* a discipline whose conclusions are more certain than those obtainable in "φυσική, *or natural Philosophy*" (*Essay* 1.3.1).[51] For Locke admitted to being so bold as to think

> that *Morality is capable of Demonstration*, as well as Mathematicks: Since the precise real Essence of the Things moral Words stand for, may be perfectly known; and so the Congruity, or Incongruity of the Things themselves, be certainly discovered, in which consists perfect Knowledge. (3.11.16)

This contention was deployed again as Locke discussed the principle that our methods of inquiry must be adapted to the nature of the ideas examined in an inquiry:

> *Morality is capable of Demonstration*, as well as Mathematicks. For the *Ideas* that Ethicks are conversant about, being all real Essences, and such as, I imagine, have a discoverable connexion and agreement one with another; so far as we can find their Habitudes and Relations, so far we shall be possessed of certain, real, and general Truths. (4.12.8)

Locke's assumption of the necessary cognitive-cum-affective use of the faculty of imagination I shall examine in this essay appears within the domain of ethics as a demonstrative science.

Excepting to a degree particularly recent imagist readings of his concept of idea, which anyway interpret him to conflate imagination and understanding, and concentrate almost solely on

---

[51] See also 3.11.15–16; 4.3.18 [see. 4.3.4]; 4.4.8; 4.12.8. In *Essay* 4.15.1, Locke remarked that demonstration exhibits "the Agreement, or Disagreement of two *Ideas*, by the intervention of one or more Proofs, which have a constant, immutable, and visible connexion one with another" (cf. 4.17.15). At 4.2.4 Locke points out that demonstrative reasoning is not casual; it requires "pains and attention" to achieve, and that there "must be more than one transient view to find" the agreement or disagreement of ideas which constitutes demonstration.

the role of imagination as the cause of an image-idea, John Locke, no less than Isaac Newton, is commonly supposed to have believed the possibility of imagination's contribution to knowledge negligible, or worse.[52] This common reading is more evident in the lack of explicit discussion than otherwise,[53] though citation of some passages in *An Essay concerning Human Understanding* and some others of Locke's writings which imply distrust or disparagement of imagination is not difficult to find.[54] One Locke scholar has noted with helpful detail just how little Locke explicitly wrote about the faculty in the *Essay*. Roland Hall then essays an explanation of Locke's apparent "disregard of imagination," concluding that the philosopher "thought he was *against* it, as a source of error and delusion, and as something that, far from leading to knowledge, prevented any regular grasp of reality."[55]

There have been some markedly ill-grounded efforts to push the pendulum in the opposite direction, notably Ernest Tuveson's and D. G. James's, but for the most part, Locke is held to have believed that there is very little if any constructive cognitive work

---

[52] Especially Michael R. Ayers, *Locke: Epistemology and Ontology* (London: Routledge, 1991), 1: 44–51, and 242–59, especially 47 and 251; see also Charles Larmore, "Scepticism," in Daniel Garber and Michael Ayers, eds., *The Cambridge History of Seventeenth-Century Philosophy* (Cambridge: Cambridge University Press, 1998), 2: 1047. There are very serious difficulties facing any reading of Locke that claims that he failed to distinguish between imagination and understanding altogether. Ayers' position is motivated largely by his interpretation of Lockean ideas as images, which is in turn supported by an historical argument in favor of Gassendi's influence on Locke on this point. For dissenting arguments, see J. R. Milton, "Locke and Gassendi: A Reappraisal," in M. A. Stewart, ed., *English Philosophy in the Age of Locke*. Oxford Studies in the History of Philosophy, vol. 3 (Oxford: Clarendon Press, 2000), 87–109; and Buickerood, "An Imagist Conception of Ideas? The Evidence of Gassendi's *Institutio Logica* and the Argument from Locke and Descartes on Geometrical Reasoning."

[53] An interesting exception to this general state of affairs is John W. Yolton, *Locke. An Introduction* (Oxford: Blackwell, 1985), 18, 132, 138, 142–45.

[54] See, for example, Donald F. Bond, "'Distrust' of Imagination in English Neo-Classicism," *Philological Quarterly* 14 (1935): 58; and idem, "The Neo-Classical Psychology of the Imagination," *English Literary History* 4 (1937): 263–64. On the whole, however, Bond was sensitive to the implications of Locke's restricted purposes in discussing imagination in the *Essay*.

[55] Roland Hall, "Some Uses of *Imagination* in the British Empiricists: A Preliminary Investigation of Locke, as Contrasted with Hume," *The Locke Newsletter* 20 (1989): 47 and 49.

for imagination, and that the most notable of its functions lead to perceptual and cognitive error, and to the delusions of enthusiasm and madness.[56] Locke's articulation of the distinction between 'real and fantastical ideas' saddles the latter with the burden of failing to conform to whatever reality of being to which they tacitly refer, or being internally inconsistent, or erroneously signified by common linguistic usage, thus laying the ground for cognitive and communication errors of numerous sorts (*Essay* 2.30). His conception of madness lays responsibility for that affliction with imagination (2.11.13; 2.33.4), and of course the specific malady that he called "association of ideas" he notoriously conceived as a species of madness (2.33). Enthusiasm he characterized as substituting for reason and revelation the "ungrounded Fancies of a Man's own Brain" as the basis of both opinion and conduct (4.19.3).[57]

The standard reading of Locke on imagination has in recent years yielded far-reaching claims about his understanding of the dangers that that power holds for any attempt to acquire knowledge of reality, create and maintain political justice, and any attempt to engage in morally acceptable behavior. Locke's perceived fear of imagination's power to undermine cognition and right conduct has been interpreted as the motivating force behind his conception both of education and the most desirable form of government. Locke's liberalism, the argument goes, is at bottom an attempt "to delimit and mold the particular expressions of the imagination."[58] The proper function of government is "to regi-

---

[56] Ernest Lee Tuveson, *The Imagination as a Means of Grace: Locke and the Aesthetics of Romanticism* (Berkeley: University of California Press, 1960); D. G. James, *The Life of Reason: Hobbes, Locke, Bolingbroke* (New York: Longmans, Green, 1949). Cf. John Yolton's review of Tuveson in *Journal of Aesthetics and Art Criticism* 20 (1961): 107–09.

[57] See Locke's journal entry for 19 February 1682 (MS. Locke, f. 6, 20–25). Locke manuscripts are held by the Bodleian Library, Oxford. References to them identify the manuscript number, followed by the page or folio citation.

[58] Uday Singh Mehta, *The Anxiety of Freedom: Imagination and Individuality in Locke's Political Thought* (Ithaca: Cornell University Press, 1992), 11. The theme of Locke's suspicions of imagination and fancy is also struck emphatically by Thomas Pangle in *The Spirit of Modern Republicanism* (Chicago: University of Chicago Press, 1988), 179–81; and by Peter C. Meyers in his *Our Only Star and Compass: Locke and the Struggle for Political Rationality* (Lanham:

ment" imagination, "to prescribe and standardize its conduct, to make it submit to conventional authority." Such a role for the state is strongly dependent upon its constitutive institutions, "most centrally, liberal education." And Locke's understanding of education "is principally a response to the volatile effects he associates with the untutored or natural imagination," to "rein in the imagination, to anchor it in the fixity of habits, to curb its potential extravagance and depth."[59]

There is much of value in this perspective on Locke's concept of mind, yet it is my contention that it exaggerates Locke's distrust and denigration of that power to the exclusion of his assumption of imagination's very real benefits for cognition and conduct.[60] Readers tend to treat Locke's view of imagination very much like they treat his view of human passions: he is read to have been unqualifiedly against each. This has resulted in an insupportably simplistic and cliché-infused conception of Locke's understanding of thought and action. What I shall argue here is that Locke held that a free agent must perform a sort of experiment *secundum imaginationem*, or thought experiment, in order to choose in accordance with his judgment. Specifically, in Locke's language, free choice entails that the agent needs to "contemplate" future goods and evils, pleasures and pains, and *by this contemplation* generate suitable uneasinesses (pains) in himself compatible with those future pleasures and pains, *and that this contemplation is a function of imagination*.[61] The free agent deliberately, voluntarily, imagines future pleasures and pains and *thereby* raises uneasinesses in himself that are suitable to them. This "contemplation" is decidedly not mere intellectual apprehension or attention; it has

---

Rowman and Littlefield, 1998), especially 125–31.

[59] Mehta, *Anxiety of Freedom*, 11 and 6.

[60] Ironically, Peter C. Meyers thinks it is Locke who has exaggerated the distinction between natural and fanciful desires, and so indicts the latter as generally dangerous to cognition and behavior. See Meyers, *Our Only Star and Compass*, 125.

[61] Some of what follows owes to Gideon Yaffe, *Liberty Worth the Name: Locke on Free Agency* (Princeton: Princeton University Press, 2001); and to my unpublished paper, "Animal Spirits and Original Sin, Appropriation and Individual Consent: Why John Locke Developed His Conception of Person in the 1680s."

an immediate effect on the agent's passions and thereby moves him to choose. There is no other way in which the free agent may motivate his free choice.

I have no desire to deny the obvious: Locke is generally believed to have repudiated any putative cognitive and moral value of imagining on the basis of very real textual evidence. Moreover, Locke's apparent antipathy for imagination and its roles in the affairs of humankind does not date only from his philosophical maturity, but goes back at least as far as his undergraduate days, when an admonition of a friend led him to observe in 1659 that

> When I complaine you conceit I accuse you, and your imagination puts a trick upon you, I can not blame you for yeelding to that which is the great commander of the world and tis Phansye that rules us all under the title of reason, this is the great guide both of the wise and the fooleish, only the former have the good lucke to light upon opinions that are most plausible or most advantageous.[62]

Thirty years later, this pessimistic conviction has quite evidently tempered to the point where Locke considered one's subordination to imagination simply a possibility, rather than a universal actuality, but it is a possibility promising dire consequences when realized:

> Thus far can the busie mind of Man carry him to a Brutality below the level of Beasts, when he quits his reason, which places him almost equal to Angels. Nor can it be otherwise in a Creature, whose thoughts are more than the Sands, and wider than the Ocean, where fancy and passion must needs run him into strange courses, if reason, which is his only Star and compass, be

---

[62] Locke to Tom [Thomas Westrowe?], 20 October 1659, Letter no. 81 in *The Correspondence of John Locke* (Oxford: Clarendon Press), 1: 123.

not that he steers by. The imagination is always restless and suggests variety of thoughts, and the will, reason being laid aside, is ready for every extravagant project. (I, §58)[63]

Locke not infrequently linked imagination and fancy with a restive mind, aimless mentation, hasty thinking, and other forms of cognitive immaturity and indiscipline. In his 1678 journal, Locke attempted to isolate means by which one might check the "extravagant…flights of imagination."[64] In *Essay* 2.1.2, fancy is depicted as "busy and boundless"; in the *Conduct of the Understanding*, the understanding, eager to "enlarge its view" by hastily generalizing and drawing inferences, is observed to enlarge only its stock of "phansies not realities" (§25).[65]

Yet the fact is that Locke did not mention imagination very much in his *Essay*, and of those few mentions, the (slight) preponderance do reveal distrust and disregard for that faculty (more frequently for "fancy") on the bases I have indicated. I shall not rehearse them here. What I do wish to emphasize is that there is also quite a—perhaps surprising—number of perfectly neutral references and allusions to imaginative power and its products, and that some of Locke's comments sometimes taken to reveal a condemnatory attitude toward that power do in fact signify nothing of the sort. The latter include remarks about imagination and mixed modes; the former are found in any number of conceptual contexts.

Despite the infrequency of Locke's direct discussion of imagination, it is clear that he acceded to the general position that imagination is the means by which we conceive nonexistent objects, properties, and events, or the means by which we conceive

[63] Quotations from Locke's *Two Treatises* owe to Locke, *Two Treatises of Government*, ed. Peter Laslett (Cambridge: Cambridge University Press, 1988). The take the conventional form of a Roman numeral denoting the treatise, followed by an Arabic numeral indicating the section referred to.

[64] MS. Locke f. 3, 20.

[65] Paul Schuurman, "*Of the Conduct of the Understanding* by John Locke" (Ph.D. diss., University of Keele, 2000), 203.

objects, properties, and events to exist in ways in which they do not in fact exist, or in ways in which we do not in fact apprehend them to exist at the time of this imagining.[66] Imagining may occur voluntarily, and the cognitive agent may deliberately imagine something should he choose to do so. On Locke's view, one perceives onself to *be* so imagining, and thus not taking the objects of imagining to exist as one so conceives them (unless mad), because one is necessarily conscious of the fact that one is so imagining, and not, say, sensing. Strictly speaking, imagining is representing a non-existent object (*pace* the above amplification) to oneself; one realizes that that object is non existent by virtue of being conscious of the act or event of imagining.[67]

What Locke's failure to discuss imagination at any length in the *Essay* compels one to do if one desires to gain some understanding of his conception of that power, is to study his earlier drafts of that work as well as his earlier journal entries mentioning that faculty. On 22 January 1678 Locke penned a lengthy entry he captioned "Memory" in which he spelled out the view of imagination he held at that time. Not merely because of the absence of any such account in the *Essay*, but also because, to the degree that any verification of this claim is possible, this 1678 account is largely consistent with what one discerns of his notion of imagination in the published book, we may, I think, take it to be the case that this journal entry expresses a position to which Locke was still committed in 1689.[68]

The entry opens with a definition of memory as reviving in the mind the idea of anything observed to exist prior to that

---

[66] See, for example, *Essay*, "Epistle to the Reader," 8; 2.2.3; 2.11.13; 2.13.24, 27; 2.17.4; 2.22.9; 2.28.19; 3.2.5, 6; 3.5.12; 3.6.46; 3.10.30, 32; 3.11.23, 24; 4.1.4; 4.4.1, 2; 4.4.15, 16; 4.5.7, 8; 4.6.8; 4.10.16; 4.11.7, 8.

[67] On Locke's conception of conscious thought, see, for example, *Essay* 2.1.19 and 2.27.9. I address the relation of consciousness and the cognitive faculties in a series of unpublished papers including "John Locke on Consciousness and Reflection," and in Buickerood, "'The Natural History of the Understanding': Locke and the Rise of Facultative Logic in the Eighteenth Century," *History and Philosophy of Logic* 6 (1985): 157–90.

[68] Considerations of space preclude such an effort here, but an unbiased survey of the discussion and use of imagination and fancy throughout Locke's writings, including his extant manuscripts, is still very much needed.

revival; to recollect the idea of an object previously perceived.[69] Then he moves on to the power of imagination. Locke initially concentrated on two of its functions: the first, when we "make a generall Idea" to represent any class in general from previous observations of diverse particulars; the second, when we join several ideas together never observed to exist together. Imagination, then, enables us to draw what Locke here called "pictures" in the mind without reference to a "patterne."[70] Whereas memory ideas always come short of their objects, imagination images do not:

> the imagination not being tied to any pattern but addeing what colours, what Ideas it pleases to its owne workmanship makes originals of its owne which are usually very bright and cleare in the minde and sometimes to that degree that they make impressions as strong and as sensible as those Ideas which come immediately by the senses from externall objects soe that the mind takes one for tother its owne imaginations for realitys, and in this (it seems to me) madnesse consists and not in the want of reason[71] ...by repeating often with vehemence of imagination the Ideas that doe belong to or may be brought in about the same thing a great many whereof the phansy is wont to furnish, those at length come to take soe deepe an impression that they all passe for cleare truths and realitys though perhaps the greatest

---

[69] MS. Locke f. 3, 16–17. See Dewhurst, *John Locke*, 100. Dewhurst's transcripts are often unreliable; and he, in any case, transcribed only parts of Locke's journals. His work must be used with caution.

[70] The account of imagination as responsible for abstract general ideas here is striking, and is absent from the *Essay* account of the origin of those ideas, which instead has them derived from abstraction. *Essay* 2.11.9.

[71] For Locke's account of madness and imagination, see also MS. Locke f. 2, 317–18 (Friday, 5 November 1677; cf. Dewhurst, *John Locke*, 89); MS. Locke f. 2, 348 (Thursday, 11 November 1677; cf. Dewhurst, *John Locke*, 90); *Essay* 2.11.13 and 2.33.4; contrast with MS. Locke f. 1, 320 (Wednesday, 15 July 1676; cf. Dewhurst, *John Locke*, 70, where "fatuitas," which means 'folly' or 'foolishness,' is mistranslated as 'madness,' rendering this entry inconsistent with Locke's position expressed elsewhere.)

part of them have at several times been supplied only by the phansy and are noething but the pure effects of imagination. This at least is the cause of great errors and mistakes amongst men even when it does not wholly unhinge the braines and put all government of the thoughts into the hands of the imagination as it some-times happens, when the Imagination by being much im-ploid and geting the mastry about any one thing usurps the dominion over all the other facultys of the minde in all other, but how this comes about or what it is gives it on such an occasion that empire how it comes thus to be let loose I confesse I cannot guesse.[72]

Imagination is by nature busy and boundless, incessantly attempting to present possibilities for the agent's consideration. In respect of its busyness and range, it has more in common with the passions than with any other cognitive capacity, and is thus intimately related to the passions so far as it is also the case that all passions require objects. This is especially the case with desire, as that passion is directed toward objects not immediately present to the agent, and so requires the representation of a sometimes yet non-existent object, available to the agent at that time only by imagining. It would then be unsurprising to find that Locke thought that an agent's imagining possible objects to be a necessary condition of experiencing desire for something hitherto unexperienced, and even that the imaginative contemplation of a possible pleasure could itself induce one to dissatisfaction with one's present state and so to desire that pleasure.[73] For dissatisfac-tion or uneasiness is at the least the most notable of the induce-ments to human action, including the action of thinking:

---

[72] MS. Locke f. 3, 19–21.
[73] I do not mean that imagining is a necessary condition for all desire. For instance, one can desire a good, a pleasure, that one associates with a past-sensed or presently-sensed object, in which cases remembering or sensing, respectively, would be necessary conditions for those desires.

The uneasiness a Man finds in himself upon the absence of any thing, whose present enjoyment carries the *Idea* of Delight with it, is that we call *Desire*, which is greater or less, as that uneasiness is more or less vehement. Where by the bye it may perhaps be of some use to remark, that the chief if not the only spur to humane Industry and Action is uneasiness. For whatever good is propos'd, if its absence carries no displeasure nor pain with it; if a Man be easie and content without it, there is no desire of it, nor endeavour after it; there is no more but a bare *Velleity*, the term used to signifie the lowest degree of Desire, and that which is next to none at all, when there is so little uneasiness in the absence of any thing, that it carries a Man no farther than some faint wishes for it, without any more effectual or vigorous use of the means to attain it. *Desire* also is stopp'd or abated by the Opinion of the impossibility or unattainableness of the good propos'd, as far as the uneasiness is cured or allay'd by that consideration. (2.20.6)

Locke's position that a moral agent's choices, his volitions, are caused by what he called "uneasiness" is notorious. In Locke's view, uneasiness is a species of pain, a sense of dissatisfaction with one's present state. This dissatisfaction has an object inasmuch as the agent is uneasy because he recognizes that he lacks something and so desires that something (for example, 2.21.31). For "whereever there is *uneasiness* there is *desire*" (2.21.39; cf. 2.7.2). Though there has long been dispute over the correct interpretation of Locke's understanding of the motivating power in choice, based in no small part upon the many changes he effected through succeeding editions of the *Essay*,[74] he had long considered the

[74] See, for example, Vere Chappell, "Locke on the Intellectual Basis of Sin," *Journal of the History of Philosophy* 32 (1994): 197–208; idem, "Locke on the Freedom of the Will," in G. A. J. Rogers, ed., *Locke's Philosophy: Content and Context* (Oxford: Clarendon Press, 1994), 101–21; Jonathan Bennett, "Locke's Philosophy of Mind," in Vere Chappell, ed., *The*

position sketched here, as his journal entry of 16 July 1676 exhibits:

> The mind finding in itself the ideas of several objects which, if enjoyed, would produce pleasure, i. e. the ideas of the several things it loves, contemplating the satisfaction which would arise to itself in the actual enjoyment or application of some one of those things it loves and the possibility or feasibleness of the present enjoyment, or doing something toward the procuring the enjoyment, of that good, observes in itself some uneasiness or trouble or displeasure till it be done, and this is what we call desire, so that desire seems to me to be a pain the mind is in till some good, whether *jucundum* or *utile*, which it judges possible and seasonable, be obtained....Desire...is nothing but a pain the mind suffers in the absence of some good...increased and varied by divers considerations.[75]

In the *Essay* itself as elsewhere, Locke recognized that one can experience numerous distinct uneasinesses simultaneously, and different choices or volitions are dictated by different uneasinesses. That is, depending on what the agent feels he lacks, he will be moved to different actions to resolve his various uneasinesses. There is no smooth parity among simultaneous uneasinesses; the most urgent at any given moment is that uneasiness that causes one to make a particular choice in order to relieve that feeling on the condition that one judges that to be feasible:

Cambridge Companion to Locke (Cambridge: Cambridge University Press, 1994), 89–114; Michael Losonsky, "John Locke on Passion, Will and Belief," *British Journal for the History of Philosophy* 4 (1996): 267–83; and Tito Magri, "Locke, Suspension of Desire, and the Remote Good," *British Journal for the History of Philosophy* 8 (2000): 55–70.
[75] MS. Locke f. 1, 339, 340. Published in Locke, *Essays on the Law of Nature*, ed. W. von Leyden (Oxford: Clarendon Press, 1954), 269, 270.

we being in this World beset with sundry *uneasinesses*, distracted with different *desires*, the next enquiry naturally will be, which of them has the precedency in determining the *will* to the next action? and to that the answer is, that ordinarily, which is the most pressing of those, that are judged capable of being then removed.... the most important and urgent *uneasiness*, we at that time feel, is that, which ordinarily determines the *will*. (2.21.40)[76]

The resolution of uneasiness is necessary for happiness, so far as happiness is the greatest pleasure and absence of pain (for example, 2.21.36). But resolving such uneasinesses and accounting for such resolution is complicated by the all-too-familiar fact that one may judge oneself to be in the position of needing to refrain from resolving one particular uneasiness now in order to relieve a greater one later. One may defer immediate pleasure in order to obtain greater future pleasure. Agents have the capacity to determine whether or not this is the case by way of "contemplating" (as Locke noted in 1676) future and greater goods, and by recognizing that to pursue that good the present absence of which causes uneasiness may interfere with attaining the greater future good. When one succeeds in doing this with any degree of clarity, one raises uneasiness in oneself due to the absence of that future good:

We are seldom at ease, and free enough from the sollicitation of our natural or adopted desires, but a constant succession of *uneasinesses* out of that stock, which natural wants, or acquired habits have heaped up, take the *will* in their turns; and no sooner is one action dispatch'd, which by such a determination of the *will* we are set upon, but another *uneasiness* is ready to set us on work. For the removing of the pains we feel, and are at

---

[76] Here I am indebted to discussion with Katherine Bradfield and her valuable forthcoming paper, "How Can Knowledge Derive Itself? Locke on the Passions, Will and Understanding."

present pressed with, being the getting out of misery, and consequently the first thing to be done in order to happiness, absent good, though thought on, confessed, and appearing to be good, not making any part of this unhappiness in its absence, is jostled out, to make way for the removal of those *uneasinesses* we feel, till due, and repeated Contemplation has brought it nearer to our Mind, given some relish of it, and raised in us some desire; which then beginning to make a part of our present *uneasiness*, stands upon fair terms with the rest, to be satisfied, and so according to its greatness, and pressure, comes in its turn to determine the *will*. (2.21.45)

Possibility, as Locke remarked in his journal, excites our desire.[77] In order to choose an absent good in the face of more immediate pains and desires, the agent must "contemplate" future greater good and thereby create sufficient appetite and desire for that good that the uneasiness of not having it overshadows the uneasiness of not having immediately pertinent goods.

What sort of operation is this "due, and repeated Contemplation"? In the published *Essay* of late 1689 and its successive editions, contemplation is first considered as one of those mental faculties that contributes some way to the acquisition of knowledge, specifically as one of the two means of retention of ideas. There Locke claimed that contemplation is "keeping the *Idea*, which is brought into" the mind "for some time actually in view." (2.10.1) Later, in his analysis of the modes of thinking, he offered

---

[77] See MS. Locke f. 1, 341; and *Essays on the Law of Nature*, 270. Cf. *Essay* 2.7.3: if delight were wholly separated from or ideas of sensation and reflection, "we should have no reason to preferr one Thought or Action, to another; Negligence, to Attention; or Motion, to Rest. And so we should neither stir our Bodies, nor employ our Minds; but let our Thoughts (if I may so call it) run a drift, without any direction or design; and suffer the *Ideas* of our Minds, like unregarded shadows, to make their appearances there, as it happen'd, without attending to them. In which state Man, however furnished with the Faculties of Understanding and Will, would be a very idle unactive Creature, and pass his time only in a lazy lethargick Dream."

some further details about that mode, contrasting it as well with some others:

> When the Mind turns its view inwards upon it self, and contemplates its own Actions, *Thinking* is the first that occurs. In it the Mind observes a great variety of Modifications, and from thence receives distinct *Ideas*. Thus the Perception, which actually accompanies, and is annexed to any impression on the Body, made by an external Object, being distinct from all other Modifications of *thinking*, furnishes the mind with a distinct *Idea*, which we call *Sensation*; which is, as it were, the actual entrance of any *Idea* into the Understanding by the Senses. The same *Idea*, when it again recurs without the operation of the like Object on the external Sensory, is *Remembrance*: If it be sought after by the mind, and with pain and endeavour found, and brought again in view, 'tis *Recollection*: If it be held there long under attentive Consideration, 'tis *Contemplation*: When *Ideas* float in our mind, without any reflection or regard of the Understanding, it is that, which the *French* call *Reverie*; our Language has scarce a name for it: When the *Ideas* that offer themselves, (for as I have observed in another place, whilst we are awake, there will always be a train of *Ideas* succeeding one another in our minds,) are taken notice of, and, as it were, registred in the Memory, it is *Attention*: When the mind with great earnestness, and of choice, fixes its view on any *Idea*, considers it on all sides, and will not be called off by the ordinary sollicitation of other *Ideas*, it is that we call *Intention*, or *Study*: Sleep, without dreaming, is rest from all these. (2.19.1)

Here contemplation obtains as the mind considers an idea for some time with attention. And this seems consonant with most of Locke's use of contemplation throughout the *Essay*. The idea contemplated may derive from any immediate source, such as memory, abstraction, judgment, or compounding, though it will ultimately owe to one of the two fountains of experience, sensation and reflection.

Locke offered yet fuller characterization of contemplation in some of the drafts of the *Essay*, which he refrained from including in the final, published version. In *Draft B* of July 1671, contemplation, as "the keepeing of those simple Ideas which from Sensation or Reflection" the mind has received, is depicted as a form of retention which,

> Either by keepeing the Idea (which is brought in to it) actualy in view, which the minde hath a power to do, for a long continuance, though it seldam happens through the importunity of other objects drawing the lookes another way, or sleepe which at once draws a curtaine over all the Ideas of the understanding, that the same Idea is for any considerable time held under consideration, This way therefor of reteineing of Ideas in the understanding may be called contemplation.[78]

This emphasis on the difficulties in contemplating, this retailing of the sorts of distractions that undermine one's efforts to engage in it, was even more marked in the subsequent draft, called *Draft C*, of 1685. There, Locke observed that contemplation is achieved

> Either by keepeing ye Idea w^{ch}: is brought into it actually in view w^{ch}: ye minde hath a pow^{er} to doe though it seldome happens y^t ye same Idea is for any Considerable time held alone in ye minde eith^{er} from ye nature of ye minde it self wherein if they be left to themselves (as in one who gets himself not to think of any thing) ye Ideas are in Continual flux. Or from ye nature of Consideration w^{ch}: consists not in one but variety of Coherent thoughts

---

[78] Locke, *Draft B*, §23, in Locke, *Drafts for the 'Essay concerning Human Understanding' and Other Philosophical Writings. Volume 1, Drafts A and B*, ed. Peter H. Nidditch and G. A. J. Rogers (Oxford: Clarendon Press, 1990), 134.

i e. variety of Ideas; or else through ye importunity of
other objects or Ideas drawing ye looking another way, or
sleepe w^{ch}: at once in most men draws a curtaine over all
ye Ideas of ye understanding. This way therefore of
retaining of Ideas in ye understanding by continued view
of any one may be called *Contemplation*.[79]

Not only is the present sensation of objects a continual threat to
the success of any attempt at contemplation, so too would the
involuntary imagination of possibilities distract one from attending to one possibility in particular. The pleasure or pain accompanying some of these ideas may induce the mind to move to attend
these other ideas, rather than that on which one is attempting to
contemplate, as Locke went on to remark in the draft chapter 23
of book 2:

When any Idea is carefully viewed & regarded by y^e
minde as it offers it selfe & this is calld *attention*. Ideas y^t
make themselves thus to be taken notice of & doe as it
were force ye minde to an observation of them are
comonly such as are new or make more then ordinary
impression upon ye sense y^t conveys y^m in or have
pleasure or pain accompanying them or are proposed to
ye minde by any passion....When wth: great Earnestnesse
& attention ye minde by its owne proper choise fixes its
view upon any Idea considers it on all sides & will not be
calld off by ye ordinary solicitation of other Ideas this I
thinke we may call *Contemplation*.

In ye several dispositions of ye minde in reference to
these three states lyes I guesse very much ye different
strength of minde Ability & Genius of different men.

Hence also we see how places retired from great
noise & light & variety of changeing objects are helpfull

[79] Locke, "An Essay Concerning humane understanding in fower books. 1685," 2.9.1. This
manuscript is preserved in the Pierpont Morgan Library of New York City.

to Contemplacion because hereby is avoided ye entrance & importunity of new Ideas w^{ch}: are apt to divert & disturbe ye minde....

Hence also we may observe ye cause why men in passion reason ill for to reasoning ye Equall Consideration of a great many Ideas is necessary but a prevailing passion will suffer none but w^t it presents to be taken notice of given y^t ye possession excludes all others & will have y^t alone regarded but this by y^e by.[80]

When the prevailing passion is vehement or even violent, passion-induced contemplation and reasoning appears to be, if not practically impossible, at least exceedingly difficult to execute. But the sort of contemplation to which Locke appealed as a thought experiment by means of which the agent grasps and motivates his will to obtain possible pleasure is not induced by vehement passion so much as it functions to develop or heighten passion, namely an uneasiness with the agent's present state and a desire for the possible pleasure as a good. I do not read Locke to have said here or elsewhere that one attempts to contemplate in the pertinent sense utterly dispassionately. Pain or pleasure of course accompanies *almost* all ideas. Yet the thought experiment contemplation of 2.21 seems clearly conceived to raise the pain of desire considerably beyond the level of velleity which may characterize those passion-accompanied ideas by means of which it is initiated. It is conceived as well to raise passion suitable to its object, passion instrumental in attaining that object as one's goal.[81]

Locke did not contend that all contemplation excites desire. Generally, contemplation is at the very least a fragile sort of attentive perception of ideas. Yet the sort of "contemplation" which constitutes a sort of thought experiment on which Locke

---

[80] *Draft C*, 2.23.7.
[81] An adequate analysis of Locke's position on the passions and the motivation of volition is far more complex than this. As Bradfield argues in her paper, "How Can Knowledge Derive Itself?" Locke also contends that one may imagine passions in addition to imagining objects of passions.

relies in his 2.21.45 depiction of motivating volition pretty clearly can not be simply a species of the perception of ideas, or the having of ideas, for that is merely thinking, no matter how lively the idea contemplated, and "bare unactive speculation" cannot by itself move the will. Some active principle is required to move the will, some means of "raising" desire. It is uneasiness alone that determines the will, because that alone is immediately present to will:

> and 'tis against the nature of things, that what is absent should operate, where it is not. It may be said, that absent good may by contemplation be brought home to the mind, and made present. The *Idea* of it indeed may be in the mind, and view'd as present there: but nothing will be in the mind as a present good, able to counter-balance the removal of any *uneasiness*, which we are under, till it raises our desire, and the *uneasiness* of that has the prevalency in determining the *will*. Till then the *Idea* in the mind of whatever good, is there only like other *Ideas*, the object of bare unactive speculation; but operates not on the will, not sets us on work   (2.21.37)

The agent has the idea of the future greater good; by contemplating that good, he heightens a passion, a desire for that good as a component of his present uneasiness, which may be assuaged by choosing and pursuing that good as a goal. My suggestion is that this notion of "contemplation" entails what can only be considered a thought experiment, though it is a thought experiment that eventuates not simply in the acquisition of knowledge, but in the acquisition of knowledge *and* in making morally adjudicable choices. For, again, this is not an entirely rational process. Apart from the fact that the sheer having of an idea of a future good is insufficient to move the will, this operation of contemplation is not primarily a sort of illation. It is, rather, an activity by which one translates one's perceptions about the greater good into passions, specifically, uneasinesses, which alone are capable of inducing one to make choices. On Locke's conception of mind, it is only imagination that is capable of enabling such a translation. What the agent does is *imagine* this future good—remember, it is something he has not experienced, and so is not an idea of sensation—and in the course of doing this, creates or intensifies his desire for it such that he is moved to acquire or achieve it. Locke's position, then, is that the imaginative contemplation of possibilities enables us to acquire knowledge of the world and to fit our morally-adjudicable behavior within that world—a far cry from a thorough repudiation of the power to generate representations of nonexistent objects.[82]

[82] My thanks to the participants of the "Science and Imagination" conference for their helpful comments on an earlier version of this essay. I am grateful also to John W. Yolton and Katherine Bradfield for their responses to a subsequent iteration of it. Finally, I am grateful to the Keeper of Western Manuscripts, Bodleian Library, for permission to quote from the Locke manuscripts preserved there; to the Cambridge University Library for permission to quote from Newton's manuscripts; and to the Pierpont Morgan Library for permission to quote from *Draft C* of Locke's *Essay*.

INDEX